# How To Make
# MULTI-BLADE FOLDING KNIVES

*Eugene Shadley & Terry Davis*

*Foreword by Tony Bose*

© 1997 by
Eugene W. Shadley and Terry A. Davis

All rights reserved. No portion of this publication may be reproduced or transmitted in any form or by any means, electronic or mechanical, including photocopy, recording, or any information storage and retrieval system, without permission in writing from the publisher, except by a reviewer who may quote brief passages in a critical article or review to be printed in a magazine or newspaper, or electronically transmitted on radio or television.

Published by

700 E. State Street • Iola, WI 54990-0001
Telephone: 715/445-2214

Please call or write for our free catalog of firearms/knives publications.
Our toll-free number to place an order or obtain a free catalog is 800-258-0929 or please use our regular business telephone 715-445-2214 for editorial comment and further information.

ISBN: 0-87341-482-9

Printed in the United States of America

**Warning:** Neither the authors nor Krause Publications assume any responsibility, directly or indirectly, for the safety of readers attempting to make their own knives following instructions in this book. Knifemaking should be approached with caution and safety in mind. Those inexperienced in the use of power tools such as grinders and buffers should take extra-special precautions in their use. As any veteran knifemaker can tell you, knifemaking can be very hazardous to your health. Be careful.

# Contents

**Foreword** .................................................................... 4

**Part I: Knifemaking with Eugene Shadley** ........................... 5

Before the Beginning ................................................................ 6

Introduction ................................................................................ 7

The Shop and Equipment .......................................................... 9

Material Considerations ............................................................ 16

The Project: A Five-Blade Sowbelly ........................................ 18

**Part II: Knifemaking with Terry Davis** ............................... 88

A Great Experience .................................................................. 89

Tools, My Favorite Subject ...................................................... 91

The Project: A Wharncliffe Whittler ...................................... 99

**A Gallery of Multi-Blades** ................................................. 183

**Glossary** .............................................................................. 189

**Addresses of Multi-Blade Makers** ..................................... 191

**List of Suppliers** ................................................................. 192

# Foreword

The first memories I have of pocket knives is when I would go with my father to the Pierson-Allen Lumber Co. in Shelburn, Indiana, when I was about six or seven years old. There was a large display of "John Primble" pocket knives distributed by the Belknap Hardware Co, in Louisville, Kentucky. When I saw it my knees got weak. In that display was a four-inch premium stockman with slant bolsters, cracked ice scales, federal shield and match striker pulls. There on the blade was etched "John Primble Finest Steel." I wanted that thing so bad my teeth hurt. I asked my father if I could have it but groceries were more important.

Every time we would go there I would stand and look at that knife and ask if I could have it. That's a terrible strain to put on a young boy.

What I did get was a burning love and desire for pocket knives. That has never left me. I've never met a pocket knife I didn't like.

From that point on I always had some kind of pocket knife with me. I learned how to sharpen and care for one at an early age. When I was a teenager I sharpened knives and spit-shined shoes for spending money,

As time went on I graduated high school and finally got a job. One of the first things I bought was a good pocket knife.

I got into pocket knife trading in my early twenties. I used to watch two old men around Hymera, Indiana (where I grew up), trade pocket knives. Their names were Edgar "Sweet Pea" Smith and Emmit Kelly. They were the best of friends until the trading started. Sometimes they would come up out of their chairs at each other like they were going to fight. When the trading ended they were friends again. It was exciting.

Old "Sweet Pea" gave me a good skinning the first time I traded with him. It took me ten years to get him back but I did. He went all over town telling everyone how I'd done him wrong. They all laughed at him.

I finally started making knives when I was twenty-six years old. I made some of the most terrible-looking things you ever saw, but I loved it. Eventually I started making one-blade slip joints about 1976 and doing pocket knife repair. The repair is where I really learned how they worked and got some good old patterns.

At some point in time I started running the NKCA shows around the country. This was an excellent place to learn about the history of pocket knives in the U.S., as well as England and Germany.

Most of the early cutlers in the U.S. came from England. The companies that were starting up would go there and entice them with better wages and working conditions. These cutlers came in the late 1800s and worked the rest of their lives here. The way pocket knives were ground changed somewhere around 1930 when a lot of the old companies went out of business because of the Depression. I've always thought this was the end of the old cutlers and their fancy way of grinding using cut swedges to get their clearances. The newer way of doing it was more cost-effective and production-minded.

Traditional multi-blade pocket knives are fascinating. They were made in all sizes from small to large in many different styles. If you've ever had a chance to really look at a fine old knife made by Joseph Rogers or George Wostenholm, or by American firms like New York Knife Co., you are truly amazed by what they did. Remember they had water power and window light. The fit and finish was quite good for a production knife under those conditions and they always worked right.

There is very little done mechanically in today's handmade knives that hasn't been done sometime in the past. Of course, the fit and finish and materials—except handle materials—are better now than they have been in the history of man.

This brings me to the two men that have written the following pages on how to make a multi-blade folder. I've always been an admirer of both Terry's and Gene's work and have a great deal of respect for both of them.

To make a multi-blade that walks and talks with blades that pass each other without rubbing is a very difficult task. To write and tell someone how to do it in detail may be even harder. As I sat and read the transcripts they both sent me before I wrote this preface I understood exactly what they were saying. From Terry's explanation of how to lay out a blade tang to Gene's explanation of how to spin a head on a pin, they went into detail.

These two fellows have paid their dues for years and now they are sharing this information for all to read. This tells you what kind of men they are and how they love their work.

Believe me folks, these are two very talented men.

Tony Bose

# Part 1
# Knifemaking with Eugene Shadley

# Before the Beginning

When people ask me what I do for a living, and I tell them that I'm a knifemaker, I often get a puzzled, lifted eyebrow, sideways look of bewilderment, followed by, "You do what?" When the discussion turns to the price of custom knives, the look becomes one of disbelief. Before the conversation ends, I sometimes hear a note of envy that I have found a way to keep body and soul together doing something that I love. Of course, there are those few who walk away shaking their heads, muttering something about the water here in Bovey, Minnesota, and its effect on an otherwise normally functioning mind.

It has been a long, if somewhat predictable, road that I've traveled in reaching this stage in my chosen profession. Long, because it actually started for me at age 13, when I wanted a handgun to use on my trapline. My dad suggested, in no uncertain terms, that if I wanted one, I would have to make it. A friend of the family, Richard Riedl, was experienced in building and using blackpowder firearms. As long as he was willing, I was ready to pester him until he taught me what I needed to know. Luckily for me, Richard was long on patience and I soon had my pistol. By then, I was hooked on working with metal and making things that worked from things that didn't work.

It also marked the start of my interest in blackpowder shooting and historical reenactment. By the time I made my first folding knife, I had made many patch knives, bowies, and assorted hunting blades. Old sawmill blades furnished raw material for the blades, while handles were either black walnut or deer antler. Most of them were pretty crude, but functional. My first folder was a single-blade lockback, encased in a buckskin pouch worn on a leather strap around the neck. I had seen one like it in a muzzleloader magazine and drew my plans based on the photo.

In 1988, at the tender age of 32, I quit a less-than-promising job in a woodworking shop to pursue knifemaking full time. My soon-to-be bride-to-be encouraged me with the words, "If it doesn't work out, you can always go back to woodworking." With that happy thought prodding me, I packed my inventory along with my high hopes and headed for my first show. In simple terms, I struck out—not one sale in the entire three-day event. Nevertheless, it was a successful show: I learned my first two lessons as a knifemaker. The first was one in humility. The second was that with the abundance of quality fixed-blade makers, all with better known names than mine, I knew that making my mark in that field would be tough sledding. So I listened as I talked with people at the show. I remember being asked about "stockman knives" and "sowbellies" and mumbling something that I hoped was intelligent about trying to make a peanut. I went home determined to find my niche in folders, especially multiple-blade folders.

I have been thankful every day since then that I didn't hit the ground running at that first show. I question whether I would have had the courage to try something different had I not been forced to take a good hard look.

# Introduction

While this book is presented as a "how to" publication, it is in fact only a couple of knifemakers' accounts of how to make two of their favorite knives. Any aspiring multi-blade folder maker must understand right up front that following our "recipes" will not assure comparable results—mine doesn't even work the same way for me every time. Neither do those who have found some measure of success with these folders all use the same techniques. It has become apparent to me that there are nearly as many ways to make a folder as there are foldermakers. Terry's and my goal in writing this book is to help the knifemaker struggling to make multi-blades find some useful information herein that will help him find his own way through it.

I think that most makers would agree that no one has ever learned his trade on his own, that learning the art of making knives is not carried on in a vacuum. Becoming a professional knifemaker is an evolving process and definitely not a go-it-alone proposition. The maker who has not questioned others for information or used ideas gathered from his peers has not become as skilled a knifemaker as he could be. (And before I get into very hot water, let me say that my decision not to clutter up the discussion with a lot of "he/she" multiple choice pronouns is in no way intended to exclude those very talented females in the profession.)

Any of us who has a question will find an answer if he will only ask. That said, I confess that I have been virtually shameless in picking the brains of my fellow makers. They all know who they are, but with their permission, I'll name just a few. R.B. Johnson was the first of many who shared with me his know-how, as well as his coffee pot. He welcomed me into his shop where I was able to observe him at work, and he answered all of the questions that I was smart enough at the time to ask. I grilled him about materials, shop equipment, technique, shows—you name it. The more I learned from him and others, the more I found that I still

- Rule Number One: NEVER be afraid to ask questions—of anyone.
- Rule Number Two: Be bold enough to test your own ideas and then charitable enough to share them.

needed to learn. I wasn't bashful. When I had questions about anything—advertising, insurance, policy, photography—I didn't hesitate to pick up the phone and call someone. Over the years, I have quizzed Frank Centofante, Ron Lake, Dave Ricke, Judy Gottage, Bob Enders, Tony Bose, Terry Davis—you get the picture.

Besides sharing information, the knifemaker tradition also encourages newcomers. During my rookie days when I was tickled spitless to get $70 for a folder, I was fortunate to be able to visit for a few hours with Buster Warenski. Along with his heartening words regarding my work, I also recall Buster's comment, "Making knives isn't going to make you rich." I have to point out that this was before the King Tut dagger.

On the flip side of the coin, I also recall my dad's incredulous, "Seventy dollars for a folder! Dream on!" Dad's been an over-the-road truck driver for about forty years; and despite his initial skepticism, he became my first distributor, peddling fillet knives, bowies and skinners all the way from Minnesota to El Paso, Texas.

My earliest folders were of the lockback design. I had made only a few when I bought the book, *How To Make Folding Knives* by Ron Lake, Frank Centofante and Wayne Clay. The book helped me better understand the mechanics of lockbacks and laid the groundwork for making other folder styles. It also provided valuable techniques which I put to use in making my first spring back folder, a 2-blade peanut with the blades at the same end. The handle was blacklip pearl, and the blades were of Chris Peterson's damascus. It was in making this knife that I first used stainless steel for the liners and bolsters. That first peanut (Serial No. 222) went to my mother, Carolyn Shadley, my earliest collector and staunchest supporter.

Besides lockbacks and spring backs, I also made a handful of linerlock folders, patterned after Michael Walker's design. Along the line, I acquired a deep appreciation for the mechanics and the diversity of the spring back style and eventually abandoned the others.

After making five of the 2-blade peanuts, I waded right into a 4-blade congress (Serial No. 285). In retrospect, it would have been logical to make a 2-blade pattern with a single blade on either end before attempting a 3- or 4-blade knife; but I have not often been accused of being particularly rational. Case in point, my first 5-blade project (Serial Nos. 425, 426, 427) involved making three 5-blade stockmans at once—66 pieces altogether. This was the first time that I recognized in myself a disturbing tendency toward masochism.

Why multi-blades? For me, a well-made multiple-blade folder is an elegant blend of beauty and functionality. With the complexity of their interrelated parts, they offer an entirely new dimension to the challenge of making a folding knife. The old patterns also have great nostalgic appeal to me and the fact that there is virtually no limit to the different patterns that may be attempted keeps me interested.

Why this book? With all of the recent interest in multi-blades, the time seems to be right. Now that it's in print, I confess that the hardest part of making the knife for this book was figuring out how to put the process in words that would be not only informative but also interesting. The most difficult consideration has been whether to use the first or second person in relating how to build the subject knife. Is "I do it this way" more or less presumptuous than the assertion that "You should do it this way?" I still don't know. Regardless of whether I have succeeded or failed in this effort to pass on what I've learned, it has been educational for me: I learned that I'm glad I'm a struggling knifemaker, and not a starving writer.

# The Shop and Equipment

My shop is an 8-foot x 24-foot area partitioned from the rest of my garage. A new free-standing shop, complete with sink and commode, is on my wish-list for 1997. While my equipment is not necessarily "state of the art," neither is it especially primitive. My shop equipment includes:

- Brown & Sharpe 5x10 surface grinder (1963 model)
- Enco table top mill
- Belt grinder (I have two Bader grinders)
- Metal-cutting bandsaw
- Dust collector
- Micro lathe
- Dremel® moto tool
- Foredom® flex shaft
- Buffer
- Choil grinder
- Large vise
- Drill press vise
- Large anvil
- 1/4 inch hand drill
- Assorted small handtools
- Particle mask/respirator
- Safety glasses
- Latex gloves for use with adhesives

The surface grinder, while certainly not essential to knifemaking, proves to be a valuable asset in the shop. I had been dissatisfied with the heavy grind marks and high cost of precision ground stock. When I learned from a friend that a 1963 model Brown & Sharpe was for sale at an affordable price, I wasted no time in checking it out. I looked, I liked, I arranged for a loan; then I loaded it on my brother-in-law's truck and hauled it 180 miles home in the middle of a March snowstorm. Once I'd arrived safely home with my 1,500-pound prize, I dropped the damn thing on its side while unloading it. Bearings lay all over the driveway. I needed to learn fast how a surface grinder goes together; unfortunately the instruction manual had disappeared years earlier so I had to wing it. Naturally, that experience doubled my anxiety as a novice operator, because every time I had any problem with it, I wondered if bouncing it on the ground had damaged something and whether I had put all the pieces back in the right order.

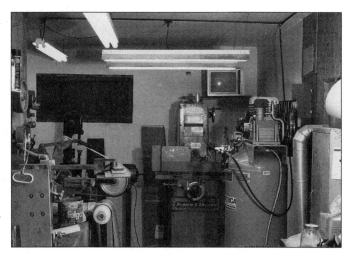

*Gene's shop is an 8' x 24' area partitioned off from his garage.*

*Gene's knifemaking equipment is not state of the art, but he has selected items well-suited to the demands of his work.*

*This Brown & Sharpe surface grinder is not essential to knifemaking, but it has proven itself very valuable for producing high-quality knives.*

# Notes On Using The Surface Grinder

One could easily devote an entire chapter, if not a whole book, to surface grinding; and trust me, what I know about surfacing grinding, you could fit in a thimble. But after tracking down information wherever I could find it, here are the high points of what I've learned about using it in my own work.

- Should you come into possession of a machine as I did, without much experience in using it, spend some time in learning as much as you can, especially about the proper coolant. Back in elementary physics we all learned that when heat is applied to one side of a metal object, the metal on that side expands, thus creating warpage. Therefore, coolant is especially important in preventing heat build-up and the resulting damage.

- Wheel dressing, i.e. running a diamond dressing tool across the face of the wheel, can produce different grit sizes, ranging from coarse to fairly fine. The surface of the chuck needs to be kept free of dings and burrs by carefully rubbing it with a flat stone. I occasionally rub a fine stone across the top between chuck top dressings. The depth of cut, width of cut and table speed are also factors in how well this machine will work (e.g. too much cut = too much heat = warpage). Different wheel dressings also play a role, i.e. a wheel of one grit size can be dressed so it will cut from fine to coarse. Depth and rate of dressing will determine this. The heavier the cut, the more coarse the grit size; a light cut results in a finer wheel grit.

- Wheel type is important since not all wheels grind all metals equally well. My personal favorite is MSC's 46H pink wheel. The cryptic letter/number labels define the grit, hardness and bond of the wheel. Incidentally, MSC is a great source of machining supplies. Its tech staff and service have been outstanding, and its catalog, with more than 300,000 items, is a valuable resource.

- As with any power equipment, safety has to be of paramount importance when using a surface grinder. During my experience, I have broken two wheels as a result of failing to heed this counsel. The wheel on my grinder turns at 2,850 rpm; at that speed, wheel fragments resemble shrapnel. I was fortunate that I was not injured and that the machine was not damaged—but it scared the liver out of me. During grinding operations, the grinder is usually left running because starting and stopping the machine can cause the wheel to get out of balance. When that happens, it must be redressed to true it up. While the wheel is being dressed, small particles fly off the wheel, creating a hazard for the operator careless enough to shun safety glasses. I also wear a respirator to keep the dust and coolant mist out of my lungs.

- My machine is equipped with a fine pole chuck which provides superior holding power. It has proven invaluable in grinding the small parts involved in folder-making.

This concludes my mini-lesson in the use of a surface grinder except for a few comments on the ideology of using power equipment in the creation of a knife. It has been said to me (as well as to other knifemakers, I'm sure), "Of course you can make nice folders. After all, you have all the tools," referring in my case to my manually operated mill and surface grinder. While the equipment is unquestionably useful, especially in maintaining the close tolerances demanded in a multi-blade knife, a machine is still only a machine, of no value by itself. If I were unable to build a folder without these tools, then owning them would not make it possible for me to do so. And a lot of folders came out of my shop before I acquired those two pieces. Like all knifemakers, I make the more sophisticated machines work for me in the same way that I make a peening hammer or a file work for me. Bottom line: A surface grinder doesn't create a knife any more than a file creates a knife—enuf said.

Knifemaking with Eugene Shadley

*Produce different grit sizes by dressing the grinding wheel; this is accomplished by running a diamond dressing tool across the face of the wheel.*

*Keep the face of the chuck free of imperfections by gently rubbing it with a flat grinding stone.*

*The platen system on this Bader grinder is drilled and tapped for two different steady rests.*

*The belt grinder's flat steady rest can be adjusted to various angles.*

## Knifemaking with Eugene Shadley

*This steady rest has a small T miter that allows you to keep the grinding material parallel to the belt.*

*A precise relationship between the steady rests and the belt is maintained with the assistance of this tiny square.*

In the shop, I also have a platen system that I put together for the Bader grinder that allows me to flat grind blades and handle material. It is drilled and tapped for two different steady rests. One is flat and can be adjusted to various angles; the other can be adjusted as well, however it has a miter slot cut in it. I made a small T miter to be used with it that is useful in grinding handle material that needs both ends parallel. When I use either of these rests, I want them to be as precise as possible. For that purpose, I use a tiny square that I made years ago. This system has proven very helpful in keeping knife joints free of gaps.

The final shop "item," possibly the most important, is an open attitude toward change. While in my personal life, I tend to be somewhat inflexible and uncomfortable with departures from the norm, my shop routine is much more spontaneous; I'm always on the lookout for innovations that will improve some aspect of my trade.

# Material Considerations

## Blade Material

For my first knives, many of which were patch knives used in loading and shooting blackpowder firearms, I used power hacksaw blades. They were good for the intended purpose, but of limited usefulness because of their variable nature, i.e. the variety of metals used in the blades. When I began making bowies and hunters, my material of choice was L-6, readily available in used sawmill blades. I found that L-6 annealed easily and heat treated with predictability because the data sheets tell exactly how to handle it. During my fillet knife madness, I used 6-inch-wide sawmill bandsaw blades, a material I could turn into knife blades without affecting the original factory tempering. When I started making folders, I looked for material that would require a minimum of surface preparation; precision-ground O-1 tool steel provided the consistent thickness and flatness that I desired and also was heat-treatable. On some of my early pieces, I also used Chris Peterson's damascus, which I liked because it was unique, like Chris.

When collectors began showing interest in my work, I knew that I needed to find a steel that would not only be good for using, but would also resist staining. I found ATS-34 to be the best answer for my purposes, because it is readily available (so far), it possesses good edge-holding ability, it offers good stain resistance, and it is easy to grind. However, heat-treating stainless steel requires more sophisticated equipment than I owned at that time. Had I not found a reliable heat-treater, I would have upgraded my equipment.

Most of my knife blades and springs start out as 3/32-inch ATS-34 sheet stock. Frames, as well as most pins, are made from 416 stainless steel, but sometimes I use sterling or gold for handle pins and shields. Spacers are of 410 or 304, and shields are of 410 unless I'm using sterling or gold.

## Grinding Belts

For about the first five years of my knifemaking career, I used only aluminum oxide belts, in the naive belief that all belts are created pretty much equal; therefore, cheaper was better since it was kinder to my wallet. Then, in a daring maneuver, I tested a 3M® micro film belt. I found that this belt doesn't twist when the humidity is high, it is long lasting and it grinds smoothly. Since that breakthrough, I have also used Norton Hoggers and 3M® apex belts.

Belts come in many grit sizes, types and backing materials. I use the tough, aggressive and long-lasting Norton Hoggers ceramic belt for rough grinding, the 80 grit for profiling parts and grinding bevels on heat-treated blades, and the 120 grit for dressing down folder backs and for further grinding the blade bevels. For finer blade grinding, I move to the apex 180 and 320 belts. Finish grinding with the 16 and then 6 apex belt leaves a surface that I can easily hand sand to 600 grit. When I work on frames and handle material, I prefer the 3M® micro film belts, which are not flexible (not floppy) and produce a flatter finish. They work well for slack-belt finishing. Because belts don't grow on trees, I want them to last as long as possible. I have found that an occasional cleaning with a belt eraser helps to prolong their usefulness.

## Handle Material

A few general remarks on the subject of handle material: For my fixed blade knives, I've used all of the common materials—is there a knifemaker out there who has not made a knife handle of deer antler? I did—no less than 50 times. I've also used horn, ivory, Micarta™, imitation mother of pearl, and various exotic and not-so-exotic woods. Once, I was accused of having killed a piano to make my wife a small letter opener with a handle of laminated ebony and ivory. The SPCMI (Society for the Prevention of Cruelty to Musical Instruments) now has me on its hit list for raiding the sacred burial grounds. I swear there was no disrespect intended—but I was born to recycle.

I now favor pearl and jigged bone for handles, primarily because traditional knife patterns seem to call for traditional materials. High-quality pearl holds a special magic for this knifemaker and good jigged bone, which is available in several patterns and colors, lends a real old-time look to the completed piece. Prime-quality jigged bone, however, can be difficult to find, but definitely worth the time spent tracking it down. More than once, I have considered learning to make my own to have a reliable source.

When I first started using pearl, its price and delicate nature were intimidating, but time and experimentation got me through the fear. I learned what worked and what didn't when using this precious material. I also learned in short order that you get what you pay for. Working around cracks and inclusions in "cheap" pearl is a tedious, frustrating nuisance. Now I search out the best pearl I can get my hands on; I can't afford to let second-rate pearl detract from the value of my folders.

Where pearl is very stable, bone is a different matter. Due to its porosity, it can absorb moisture or dry out, causing warpage, swelling, cracking and shrinking. Also, the density of bone varies, which causes the dye to penetrate differently. Two pieces of bone dyed at the same time in the same solution could easily end up different shades of the same color. Therefore, finding pairs of matching bone can be a challenge.

Proper storage of bone is important. I keep my supply tightly covered in a coffee can with a plastic lid. I believe that this helps to minimize the effects of Minnesota's humidity. Also, I apply lemon oil to the handle after the knife is completed, which seems to help prevent moisture loss/absorption. With all its quirks, I still like nice jigged bone, prefering the old jigging patterns with their random designs and pleasing colors.

## Design Develpment

When I began making 5-blade folders, I had been making 3-blade knives in the stockman, sowbelly and Congress whittler patterns. My first 3-blade stockman (Serial No. 391) came out of an assortment of old knives that I had acquired, mostly by sifting through "junk" boxes at knife shows. Keeping the best, losing the rest, I devised a pattern that I liked and that worked well. I selected blade shapes and incorporated design ideas from various photos that I had seen. My goal was to produce a knife that maintained the essence of the old original pattern without being an exact replica.

Having gained some confidence with the 3-blade patterns, I knew the time had come to test my mettle with 5-bladers. In the simplest terms, making a 5-blade folder would be as easy as adding another spring and spacer plus two more blades to a 3-blade folder. Right? The frame and basic geometry remains constant, and the master blade doesn't change. I convinced myself that if I could produce a quality 3-blade stockman, I could produce an acceptable 5-blade model.

In roughly eighteen months, I made a dozen 5-blade folders in the stockman pattern, before moving on to the 5-blade sowbelly, the subject of my contribution to this book.

Knifemaking with Eugene Shadley

# The Project: A Five-Blade Sowbelly (Serial No. 587)

Let's start this project with a brief overview. As indicated, this knife has 22 pieces, all of which must fit together exactly. The knife will have blades of ATS-34, frames of 416 stainless, pins of 416 and spacers of 410. It will sport jigged-bone handles and a shield of 410 stainless. The accompanying sketches, created by Jim Corrado, based on tracings of my patterns, will help identify the various parts.

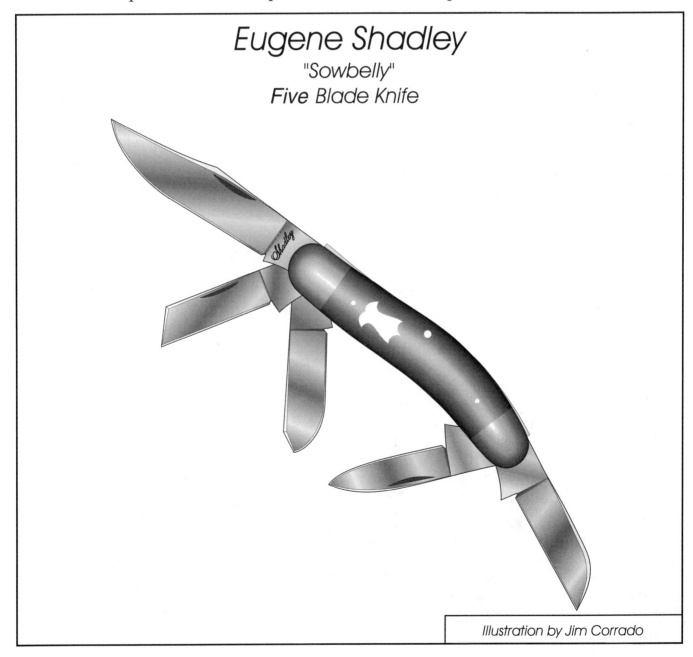

Eugene Shadley
"Sowbelly"
Five Blade Knife

*Illustration by Jim Corrado*

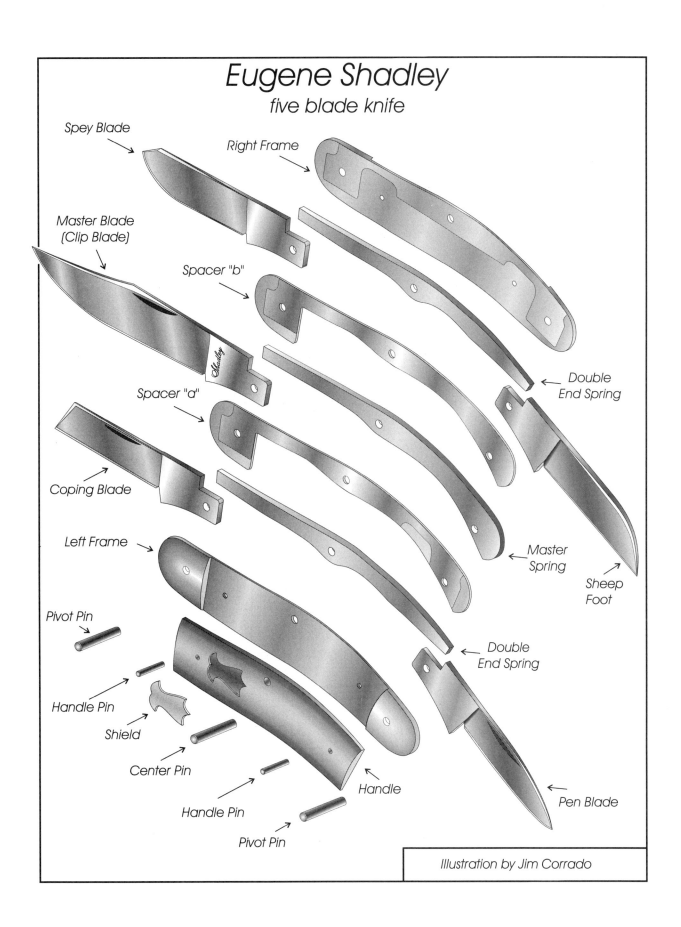

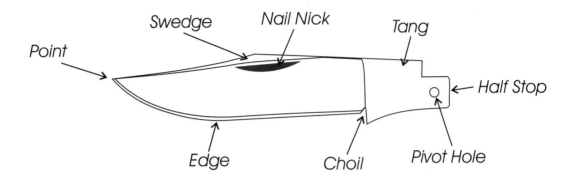

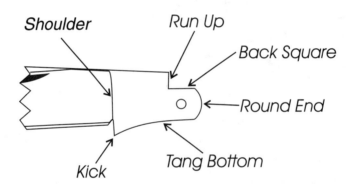

The authors have chosen to use the above nomenclature to define these parts, although others in the knife world may use other terminology.

*An Enco metal-cutting bandsaw, modified to allow the user to run it comfortably in a standing position.*

## Cutting The Parts

Next to traveling, cutting parts is probably my least favorite job. I will do whatever I can to ease the pain. For equipment, I use the Enco horizontal/vertical metal-cutting bandsaw. I made some modifications to mine to comfortably run it in a standing position. The Enco has three blade speeds, and I use the highest speed setting, 200 FPM, which is still very slow, compared to a wood-cutting saw. There are three things required in using this machine safely: a push stick, safety glasses and a high level of concentration. Even though it cuts relatively slowly, *this machine can hurt you.*

I like and use MSC's Starrett Matrix II power-band blades with 24 cobalt teeth per inch, which helps to cut down on tooth loss and avoids jamming the blade. Cobalt is a lot tougher and longer-lasting than carbon steel.

For patterns, I normally use mild steel, because I use the patterns only to scribe around and to locate pin holes. With this limited wear and tear, it takes a long time to wear them out. Obviously, the better the pattern, the better the resulting product.

When tracing around the patterns, I scribe as near to the pattern as possible and then cut out the parts a little oversize, which gives me latitude when fitting parts. There will be plenty of opportunities to grind a little off, but not even one chance to add any back. When I grind the parts for profile, I grind almost to the line, allowing a little slack for adjusting the parts when they return from heat-treating.

After grinding parts to profile (to 80 grit), I deburr the edges by rubbing the parts on a sanding plate, which is simply a piece of sandpaper glued to a surface-ground plate and clamped in the vise. (This sanding plate will be used often throughout the project.) I check the parts for flatness, making corrections by placing the part, bowed side up, over the round hole in my anvil. Just a tap or two with a small hammer usually does the trick.

Once they are reasonably flat, the parts get grouped according to knife type. This provides a visual check to assure that all the parts for each pattern are of the same thickness and that I haven't miscounted. While corrections can be made later, it simplifies things when they're all at the same thickness. At this point, I surface-grind the parts just enough to clean them

*Scribe to your pattern as closely as possible, then cut the parts a little oversize to give yourself some latitude in fitting them together.*

*A cut-out blade showing the extra clearance around the traced lines. It's better to have extra material to grind off, because if the blade starts out too small you can't make it bigger.*

*Grind the parts to profile (to 80 grit) almost to the scribed line, allowing some room for adjusting size after the parts return from heat-treating.*

*The belt grinder's steady rest is an invaluable tool, but caution must be exercised in using it: Minimize the gap between the belt and the rest or the part can get caught between them.*

*Deburr the edges of the ground blades by rubbing them across a sanding plate, which is a piece of sandpaper glued to a ground plate.*

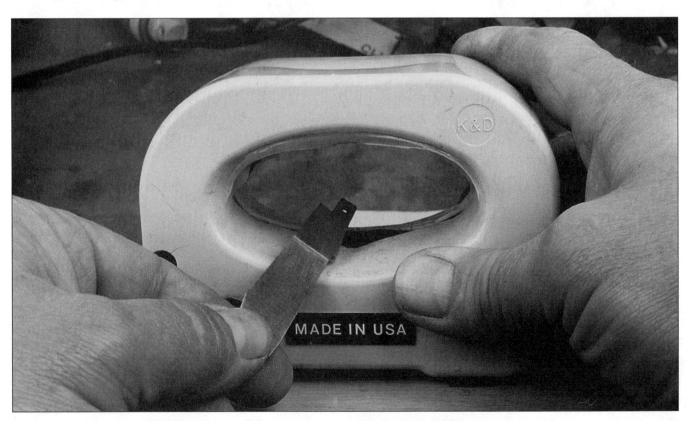

*Putting all the ground parts through a demagnetizer helps keep metal fragments from sticking to them.*

*Using a milling machine fitted with a drilling block ensures precision when drilling holes through knife parts.*

*After drilling the holes, it's time to crink the blades to the desired angle using this device, the crinking tool.*

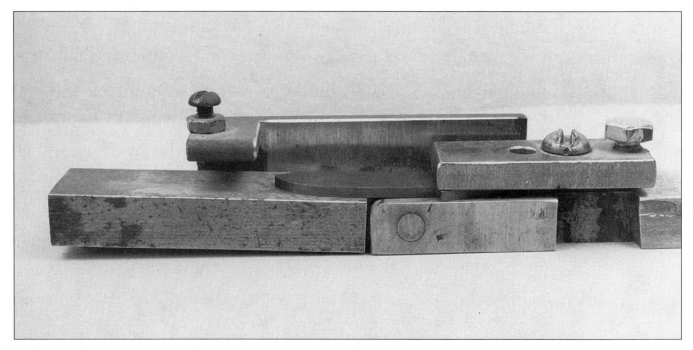

*Here, one of the blades is bent to a preset angle in the crinking tool.*

up. This puts a slight burr on the edges again, so I clean up the profiles with a 220 belt on my Bader grinder.

Next, I rub the square edges with a crock stick, a round ceramic rod commonly found in knife sharpening kits, removing the sharp edge. The parts are then put through a demagnetizer, which helps keep metal fragments from sticking to the parts. (While not an absolutely necessary step, this may help preserve your sanity.)

At this point, I lay the patterns over the parts to locate and mark the holes for drilling. I prefer the milling machine over a drill press to drill the holes, because the milling machine has bearings designed to prevent spindle wobble. This is important in obtaining as precise a hole as possible. I have attached a small block to the table that provides a drilling platform that can be machined flat and perpendicular to the spindle, affording additional insurance of precision drilling. Another advantage to the milling machine is that the table is not likely to get knocked out of adjustment.

Using bits of cobalt split-point, again from MSC, I first drill all the center-punched holes with a 1/16-inch bit. Then I switch to an .089-inch bit, finishing with a .0945-inch reamer. It's a good idea to always use sharp bits for easier drilling. The burrs left from drilling and reaming need to be removed on the sanding block before sending out for heat-treating block. Now is also the time for any necessary crinking. I use a tool that bends the blades to a preset angle. The parts are ready to be degreased, packaged and sent to the heat-treater.

- Safety tip: I have to be careful when grinding while using the steady rest. If the gap between the belt and the rest is too great, the part can become caught and break the belt or cause personal injury.

## The Frames and Spacers

The frames are an important component of the multi-blade folder. They not only keep the blades safe when not in use, they also keep them in alignment, preventing them from rubbing against each other. I use 416 stainless, which has characteristics I like. Besides being stain-resistant, it has very nice machining properties, in that it drills and mills easily.

In making the frames, I first cut the 1/8-inch-thick bar-stock to length and deburr the ends, checking the parts for flatness and straightening them as necessary. Next, all of the frame parts get surface-ground and demagnetized. Then I lay the patterns over the parts and mark the pivot holes for drilling. I use 1/16-inch bits to drill the holes. I reapply the patterns to scribe the frames to shape. Using the milling machine, I cut the interior relief areas. With a moto-tool, I clean them up with rubber aluminum oxide wheels that can be obtained from a dental-equipment supplier.

Next, I turn my attention to the handle area, taking it down to .040 inch on the milling machine. I set stops on each end of the cut to leave .010 inch for a smooth final cut on the inside of each bolster. The frame halves are machined one at a time on a plate that I fabricated to hold the parts. The plate is designed to provide for varying positions. Alignment pins and clamps secure the part for machining. After cutting the front of the frame, I loosen the clamps and move the rear alignment pin to its second position. The reminder of the excess material can now be removed.

That done, we're ready to profile the frames, cutting them on the band saw and grinding to profile on the Bader. Next, I deburr and flatten the parts, using an attachment that I've set up on my anvil. I also flex the parts by hand and check them for flatness against the back of my dial caliper. Once I'm satisfied, I pair them up and put alignment pins in the pivot holes. To hold them together for drilling

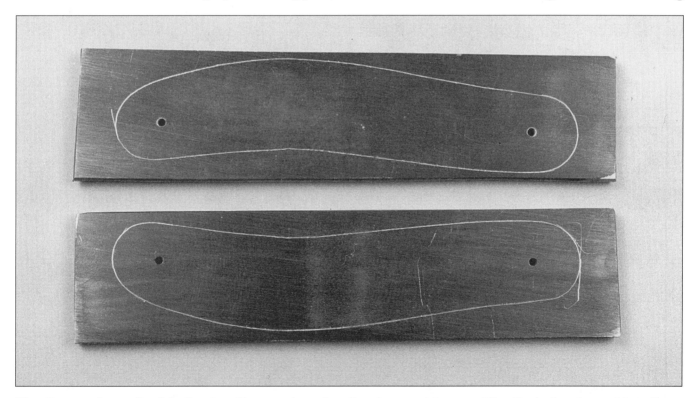

*The frames keep the blades in alignment and safe when not in use. The first step in making them is to scribe their shape (using a pattern) on 416 stainless steel barstock.*

*Use a milling machine to cut the interior relief areas of the frames.*

*Clean the frame interior relief area using a moto-tool fitted with a rubber aluminum oxide wheel.*

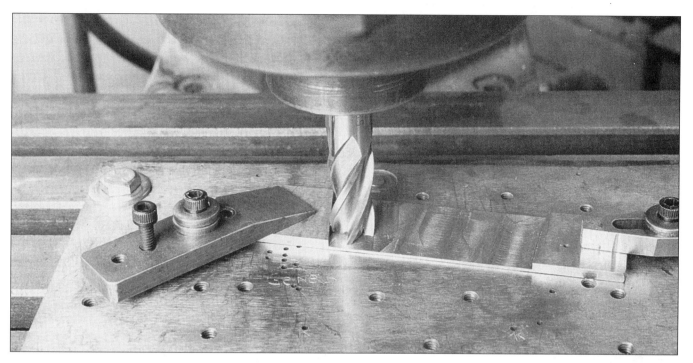

*Machine the frame halves one at a time; mill the handle area down to a thickness of .040". This special plate, fitted with pins and clamps, allows alignment in a variety of positions.*

*The spacers are cut the same shape as the frames, then the relief area is cut away using the milling machine.*

and reaming, I use specially made clamps. Once the clamps are in place, the pins are removed and the frames drilled and reamed. This assures that the holes are true and in alignment (if they're not, it's "bald city" later on). The center spring-hole placement is determined later when I'm ready to trial fit.

Next, I lay out the frames, springs and blades, putting them in groups. For instance, a five-blader will have one set of frames, three springs, five blades and two spacers. I prefer to select the handle material at this time for two reasons: so that I know that I have suitable material on hand and just to have the decision behind me.

Finished for now with the frames, I move on to the spacers. Usually, I cut them from .040-inch 410 stainless, although for jack patterns (both blades at the same end), I prefer .030 inch because I like a slimmer profile. I scribe and cut the material for the spacers in the same shape as the frames and cut the relief areas on the sides. Now, the parts can be drilled to the proper-size pin holes. As with the frame-spring pin holes, the spring center pinhole is drilled later.

## The First Trial Run

This process involves grinding springs and tangs to fit. As mentioned, all parts have been cut slightly oversize. After repositioning the patterns on the blades and springs, I rescribe them. When I grind the tangs and springs on a five blader, I group the parts together relative to the way they fit together in the folder. There are three blades on the master end and two blades on the other. The two small blades from the master end are glued (with Super Glue™) to either side of the master blade. Then I glue a tang-size piece of Plexiglas™ on the bottom side of this stack. Since the small blades are crinked, they cannot lay flat on the steady rest for grinding the tang outline; the Plexiglas™ compensates for this. I place the cope blade pattern on the stack, using an alignment pin, then scribe the tang portion. Using a 120-grit belt, I grind just about to the line, before moving on to 60-micron, 30-micron and 15-micron belts. I polish the top portion of the tangs of all three blades, using a sewn cotton wheel with green-chrome rouge.

Leaving the Plexiglas™ block in place, I remove the master blade from this stack, and scrape the glue off the blades. After regluing the two small blades together, I grind and polish the rest of the tang. Then I finish grinding and polishing the tang on

*In preparing to grind the tangs to fit, stack the blades the way they will fit in the knife, Super Glue™ them together, then attach a Plexiglas™ piece to compensate for the crinked blades.*

*The assembly block, fitted with center punch and pivot end pins, is ready to mark the left frame.*

*The left frame is the first part placed on the assembly block.*

the master blade. This process must be done in two steps, since the kick on the master blade is different from the two smaller blades (it must stick up above the others when the folder is closed). I repeat the process on the other end, again gluing the two blades together and using a Plexiglas™ piece on the bottom, as before. The two blades on this end can have the entire tang ground and polished simultaneously, since they have identical geometry.

The two double-end springs get paired up and pared down to their final inside dimensions. I spot-glue the two springs together, scribe and grind them, starting with 120 grit, down to 15 micron. I rescribe the master-blade spring and grind it also.

Next, I put the left frame on an assembly block. This block has a 3/32-inch hole, 1/4-inch deep, located where the center pin needs to go. I've made a small center punch out of a 3/32-inch drill bit to go into this hole. The pivot end pins are put in the block and the left-hand frame is put on these end pins. The frame liner is now laying on the center punch. I carefully tap the liner with a small hammer, thus marking the hole for drilling. Moving to the mill, I drill the frame with a 1/16-inch bit. Then I take a couple of 3/32-inch pins, flattened on one end, and put the two frames together, with the spacers in between them. The drill is run through the first 1/16-inch hole, drilling through the spacers and the second frame. Finally, I remove the 1/16-inch bit and install a .0945-inch bit and collet, re-drilling the hole. At this point, all blade and spring holes have been completed.

Returning to the assembly block, I remove the center punch, replacing it with an assembly pin. The left frame is put on the block and a double-end spring is put on the center pin. I indicate which blade belongs on which end of which spring by means of scribed notes on the parts. The springs have been left long enough to allow for adjustment. I position the cope blade on the right-hand pin and position the blade against the spring, scribing the end of the spring. I take the spring to the grinder and shorten it, using 60-micron through 15-micron grits. I must take care to avoid over-shortening the spring or the blades will not line up correctly. Now it's back to the block to test the fit. If it is still too long, it goes back to the grinder where I use 30-micron, followed by 15-micron belts.

Now the pen blade goes on the left-hand pin, and I repeat the process. Once the spring is acceptable (the blade is in its proper position in relation to the back of the knife), I put the cope and pen blades together on the block with the spring in place. I made a tool from a piece of 3/16-inch bronze rod and a file-handle that I use to help push the spring up for assembly. Next, I check the half stops, again being careful not to take off too much, just getting them close. I check the kicks and tang bottoms. I don't get too carried away at this point with precision—fine-tuning can be done later, and I don't want to have to make another part if I find I have taken off too much too soon. If the spring seems unusually strong, I can reduce the tension by thinning the spring and grinding a little off the belly (keeping in mind that the tops of the springs will be ground flush with the top of the frames and liners). If I take off too much now, the springs may become too weak.

Next, the cope, pen and spring are taken off the assembly block, leaving the frame intact. This keeps the pins solid and provides a reference location for positioning the blade. The master blade and spring are fit in the same manner, then removed. I repeat the process (used in fitting the cope and pen blades) with the sheep and spey blades, again removing the parts from the frame when finished.

*The pen blade is placed on the left pin, the cope blade is placed on the right pin and the spring is fitted in between.*

*This tool, made from a piece of bronze rod and a file handle, is used to help push springs up for assembly.*

*Here, the master blade and spring have been added to the assembly block.*

*Using the spring tool, push the spring to provide clearance to fit the second blade in each pair.*

*The sheep and spey blades and their spring are the last parts fitted on the assembly block.*

*Scribe the spacer to define the area that needs to be trimmed to allow the blades to fit inside the folder.*

*After the parts have been assembled, dress down the back of the knife on an 80 grit belt.*

Once this initial fitting process is complete, the entire folder can be assembled. First, I put the cope and pen blades on the frame, followed by spacer A; then the master spring and blade are positioned over spacer A. The spacers on this folder need to be trimmed to allow room for the blades to fit on the inside of the folder. I define the area to be trimmed by means of a scribed line on the spacer around the spring and tang area of the master blade. (The actual trimming comes later.) Next, spacer B is put in place, followed by the spey and sheepsfoot blades and their respective spring; finally the right frame is installed.

If I haven't made any mistakes up to this point, I can dress down the back on an 80-grit belt. Note: Because of the unique shape of the sowbelly, the pen and sheep blades must be at half-stop during this process. All springs and blades will be flush in the open position. No magic here—if I've ground them flush, they'll still be flush.

However (and this does sometimes require magic), the springs must remain flush when the blades are at the half stop and closed positions. One at a time, I check each blade, making notes on paper of necessary tang adjustments (for instance, I may need to take some off of the half-stop). The springs should stick up slightly or be flush in both positions. If they fall below the tops of the liners and frames, I've taken off too much material in an earlier process. Sometimes, this miscalculation can be corrected by taking a little more off the half stop and rear top tang, *a little at a time.* If too much comes off, the spring tension will be decreased to

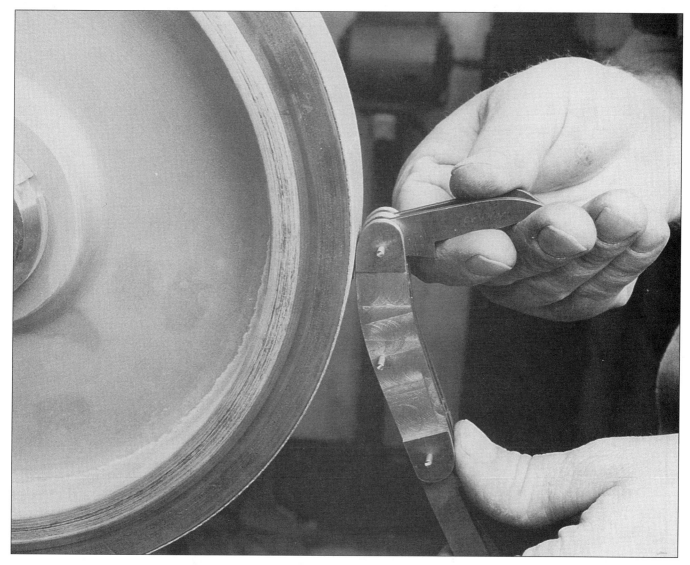
*Note that the pen and sheep blades must be at half stop while dressing down the back of the knife.*

the point where it will not match the rest of the blades. If that happens, the only option is making a new blade.

On the other hand, if the springs stick up above the back of the knife, the tangs need to be adjusted. I disassemble the knife by pulling the pins from either end of the frames—never the center pin. (There is very little material holding the center pin, and it will not withstand repeated extractions.) I refer to my notes in making the necessary corrections, grinding and refitting the blades in the same manner and order as during the initial fitting. I sometimes have to repeat the process four or five times or more before I'm satisfied with the fit. When I've reached that point, I redress the backs with a 120-grit belt.

Now that everything is working properly, I need to make sure that the blades and springs are all the same thickness. I start by disassembling the folder. But before dealing with the blade and spring thickness, I want to trim the spacers to their final shape. Using the bandsaw, I remove the excess material on the scribed spacer, cutting close to, but not on the line. Using the first one as a pattern, I scribe the second spacer and trim it. After deburring the edges, I spot-glue the spacers together with the master-blade

Knifemaking with Eugene Shadley

Trim the excess material off the spacer by cutting along the scribed lines with a bandsaw. Note that you should cut near, but not on, the line.

Grind the spacers, removing the excess material right up to the scribed line.

spring sandwiched in-between, inserting 3/32-inch pins in the holes and attaching an ordinary office binder clip to assure correct alignment.

After the glue sets up, I remove the pins and clip. Then I move to the grinder, removing the remaining excess to the line, finishing by hand-sanding to 600 grit. Note that grinding in the sandwich mode is limited to the area between the center spring hole and the rounded end of the master spring. I remove the spring from the stack by soaking in acetone. Then I glue the spacers together again and grind the remaining inside surface. This sandwich method helps ensure that the inside cutout areas are ground exactly the same. If done properly, the spacers and master spring will be flush on the inside of the folder, providing a clean, crisp appearance. Finally, I separate the completed spacers by soaking in acetone.

## Grinding Blades and Springs to Thickness

For grinding blades and springs to thickness, I begin by making sure that the chuck is free of burrs. I apply the dressing tool to the chuck and activate the magnet. Next, I bring the dressing tool into position and lock the table. The wheel (46 grit) is dressed for a fine finish, the dressing tool

- Note: Surface grinding blades that have been crinked presents a special problem because the pieces cannot be laid flat on the chuck. My answer to this has been to let the blade portion hang over the edge with the tang flat on the chuck. This type of grinding demands great care—it is not a procedure to be attempted by the beginner.

is removed from the chuck, and the chuck is cleaned of accumulated abrasive dust.

I rub the spring sides and blade tangs on the sanding block to remove the burrs that occurred when dressing down the back of the folder, and I check the parts for flatness on one side. I put a piece of flat stock, 1/16 inch x 1-1/2 inches x 5 inches, on the chuck against the stop. This serves as a backing plate so the parts don't slide back on the non-magnetic part of the chuck. The plate must be thinner than the parts that are being ground, so that the wheel may pass over it without touching.

I then place the flat side of each spring down on the chuck with the springs positioned against the backing plate and activate the chuck. Using lacquer thinner, I clean the area of the chuck where the blade tangs will be placed. The lacquer thinner is important because we'll be using Super Glue™ to spot-glue the tangs, and the surface must be free of all residue.

Once everything is in place, I spot the tangs with Super Glue™ on the upper left-hand area of each tang, where the tang and the chuck make contact; this keeps them from sliding around on the chuck. I take care not to get any glue on the tang flat. I spray the tangs with Zip-Kicker™ accelerator, wiping off the excess.

The blades and springs are now ready to be ground. First turning on the coolant, I grind the parts, taking off about .0002

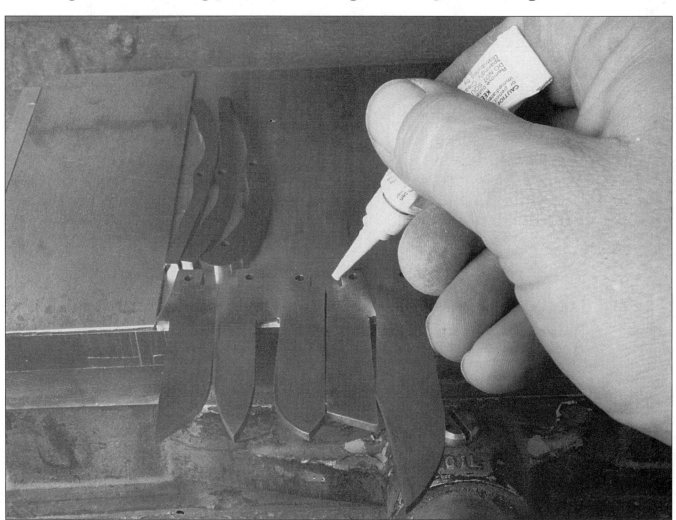

*Before grinding, spot glue the tangs to keep the blades from sliding around on the chuck.*

*Take care to stop the infeed at the same place with each pass of the grinding wheel, or you'll risk hitting the higher unground portion of the blade.*

inch with each pass. (For reference, any one of my few remaining hairs measures .002 inch in thickness, ten times the depth of the cut.) I must take care to stop the in-feed at the same place each pass. If I overshoot the runway on this, I could easily knock the piece off the chuck by crashing the wheel into the higher, unground portion of the blade. Once I've ground the blades to the prescribed thickness, I release the magnet and pop them off, rinsing off the coolant and blowing off the blades with my air hose.

I drop the blades into a jar of acetone to remove the Super Glue™ and remove the backing plate from the chuck. Next, wiping the plate and chuck with a small rubber squeegee, I remove the Super Glue™ from the chuck with a brass scraper and clean everything again with a fresh rag. I'm now ready to repeat the whole process on the flip side—second verse, same as the first. When all the old (pre-heat treat) grind marks are gone, the parts are again taken off the surface grinder.

The tangs need no further grinding, but the springs are not done yet. They must be relieved so that the tang ends move freely. Using a magnetic parallel block that has been ground at a slight angle, I place the springs on the block and grind from one end almost to the center hole. Then I flip the spring over and, putting a .0005-inch shim under the front, I grind again. The master spring will only need this done on one end. However, the double-end springs

*Relief-grind the springs, so that the tang ends move freely, using a magnetic parallel block that has been ground with a slight angle.*

*To finish the springs, rub down the inside surface with very fine Scotch-Brite™ pad.*

will need to be turned around, and this step repeated on the other end.

The springs are finished for now, except for rubbing down the inside surface with a very fine (gray) Scotch-Brite™ pad. I hand-sand the blade tang sides (flats) with 600-grit paper, wrapped around a Plexiglas™ block, rubbing the parts in a motion perpendicular to the surface-grinder marks. This allows me to see when all the surface-grinding marks are gone. When this is completed on all blades, the tang holes need to be cleaned—a pipe cleaner works fine. The folder is ready for the second assembly.

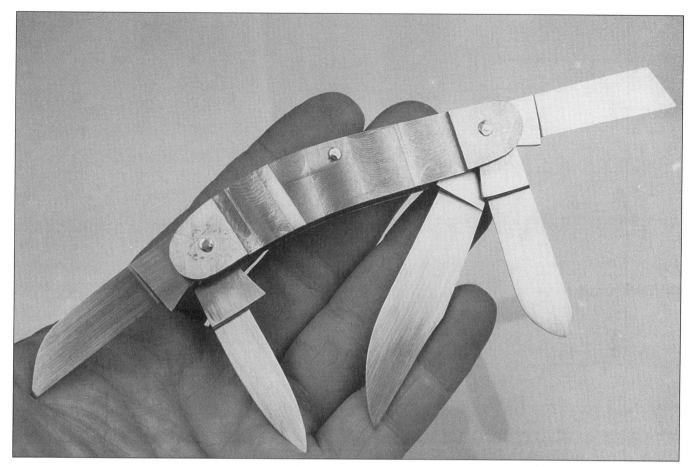

*The second assembly of the five-blade sowbelly after grinding blade bevels.*

## Second Assembly

First using an air hose to blow off the parts, I assemble the folder, trimming and peening the pins with a ball-peen hammer. The raised heads on the pins will be easy to grind off later for the next disassembly.

With the folder reassembled, the blade sides may now be ground to their proper blade tapers. Grinding the proper angles on the sheep, pen, cope and spey blades is very important. The blades must be ground so they pass by each other without touching. They also need to be left thick enough so they don't flex too much and rub when opened. Crinking can help this. However, it's not the whole answer; proper grinding is also important.

Starting with the master blade, I grind all blades on an 80-grit Norton Hogger belt. During this process, I use a tool fashioned from a chainsaw file handle and a 3/16-inch bronze rod. The end of the rod has a step cut in it on which to rest the blade. The fact that I still have fingerprints indicates the effectiveness of this device.

The flat platen I use for a backing to do this grinding is 1/8-inch x 2-inch x 6-inch heat-treated ATS-34, left hard. I use two drops of Super Glue™ to hold it in place while I do the grinding. The platen is popped off when not in use. Every so often, I surface-grind the plate to true it up.

After the master is roughed out, I move to the pen or sheep-foot blades. I grind them first on the liner side, because I want the blade to close without touching the liner. Once I'm satisfied with the clearance, I grind the other side of the blade. I have to remember—not too thin, or there

*Trim the pins and peen with a ball peen hammer. The pins' raised heads will be simple to grind off for the next disassembly.*

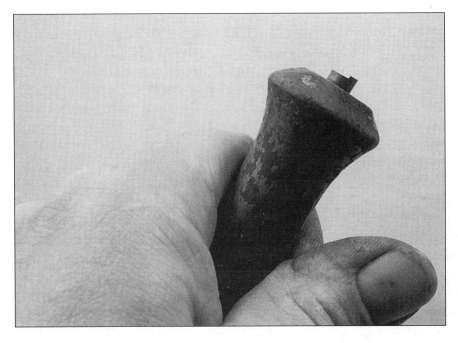

*This tool, constructed from a chainsaw file handle and a bronze rod, is used to hold the blade during grinding.*

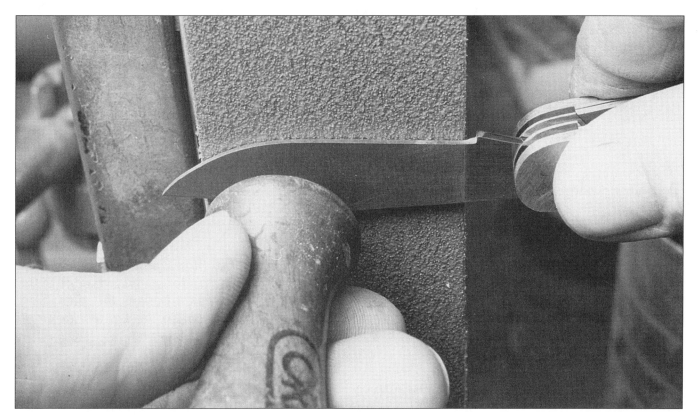

*Rough-grinding the master blade bevel.*

*Flip the blade around and grind the other side using the same technique.*

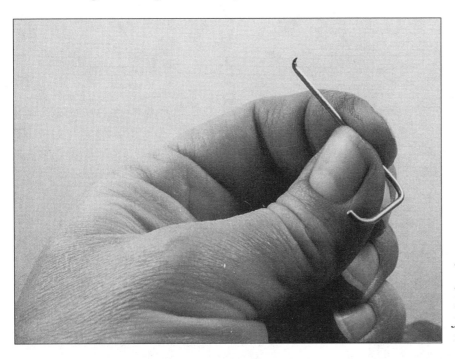

*This blade puller is a tool that slips under the cutting edge of the blades and opens them up—useful because the blades lack nail nicks at this point.*

will be too much flex. Also, at this stage, it can be hard to get the blades open, since they don't have nail nicks. Pliers will work, but I use a blade puller that I made. The tool is used under the blade on what will be the cutting edge.

After the sheep and pen blades are ground, it is time to rough out the cope and spey blades, again being careful not to get carried away on the grinder. When all blades can be closed at the same time, they are ready to be run through the grit sizes. I use a 120 Norton Hogger belt to clean up the 80-grit marks, then switch to apex belts, starting with 320, through 400, 16 and 6 sizes. Then the blades are hand-sanded to 600-grit.

After the blades are hand-sanded, the swedges are ground on the blades. Besides adding to the aesthetic appeal of the knife, the swedges give the blades a little more clearance. I first grind the swedges on a Hermes RB 406 Flex P400 belt, then switch to a 9-micron belt. To finish, the swedges are hand-sanded to 1,200 grit.

Nail nicks are next. I use my milling machine, clamping the folder on a fixture I made for the purpose. By loosening a set screw, I can tilt the folder to whatever angle I need. I use a white tool post grinding wheel that has been dressed to the correct angle.

The nick locations are marked on the blades with a magic marker. With the folder clamped in the fixture and one blade open and in position, I move the wheel into position and start the mill. I turn the in-feed handle and slowly adjust the wheel so that it just makes contact. Holding an air hose in my free hand to blow air on the blade, helping to keep it cool, I turn the handle in slowly. On the small blades, I grind to a depth of .035 inch; the master blade is ground to .045 inch. When all nicks are done, I take the folder to the buffer and buff the upper edge of each nick, thereby softening its top (helps save thumbnails from chipping). Then, with a slip of 1,000-grit paper and a tapered Plexiglas™ block, I hand-sand the inside of each nick. Next, the blade sides are sanded with a sandpaper-backed Plexiglas™ block. Already sanded to 600 grit, they are now brought to 800, 1,000 and 1,200.

Now, for the next disassembly. Starting with the master-blade end, I grind off the

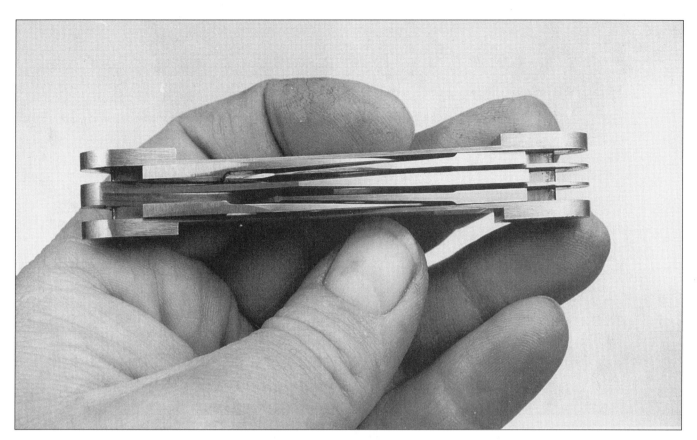

*The blades are ground until they can all be folded closed at the same time.*

*Hand-sand the blades to 600 grit Klingspor.*

### Knifemaking with Eugene Shadley

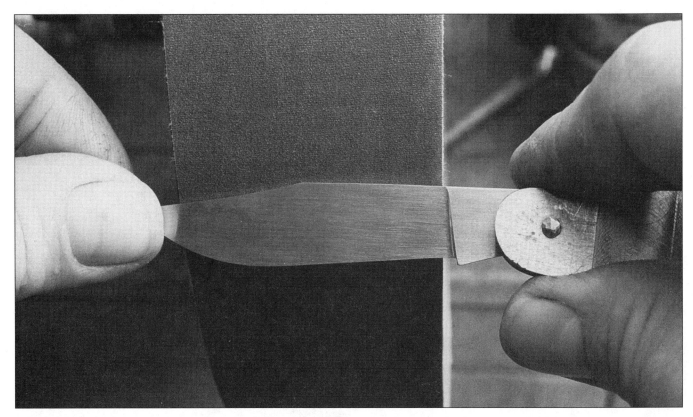

*After hand-sanding, grind the swedges to increase the aesthetic appeal and provide more clearance.*

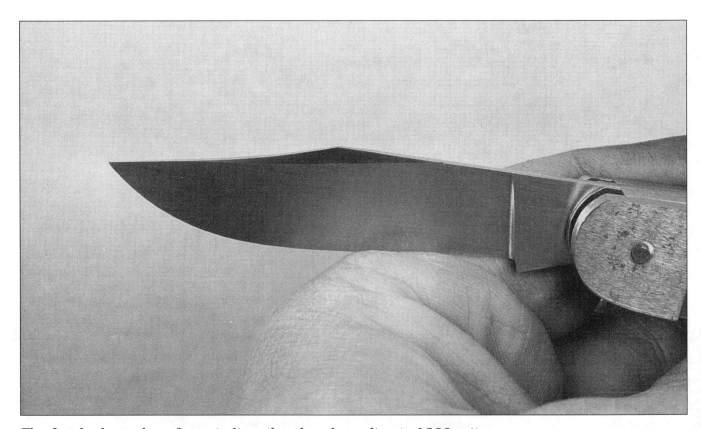

*The finished swedge after grinding, then hand-sanding to 1200 grit.*

*The milling machine is fitted with a nick fixture in preparation for grinding the nail nicks.*

*Mark the nail nick locations on the blades with a magic marker.*

*With the blade held in position, slowly turn the infeed handle until the nick fixture makes contact.*

domed heads of the pins. With the folder positioned so that the pin to be punched is placed over a corresponding hole on the anvil, I drive out the pin with a punch and hammer. After all three pins are extracted, we're ready to move on to the next step—grinding the choils.

For this procedure, I made a special grinder by mounting a 3/4-hp, 3,450-rpm motor on a frame. I made the rest with a miter-groove cut in it, milling the miter slide from 3/8-inch square stock and attaching it to a piece of 3/16-inch aircraft aluminum plate. To steady the tang, I use a small rubber-tipped hold-down clamp, placing the tang under the clamp and aligning it with the wheel. The whole miter assembly is pushed forward against the wheel and the cut is made. For the wheel, which needs to be dressed to the shape of the choil and needs to hold its shape, I use a 1/4-inch x 7-1/4-inch surface-grinding wheel. This has worked very well for me. After grinding the choils, I polish them with a Cratex wheel shaped to fit in the ground-out area. The blade tang flats are then hand-sanded with 800-1,200-grit Klingspor paper, using a Plexiglas™ block.

Now I'm ready to etch the master blade with my mark and with the next consecutive serial number. First, cleaning the part with lacquer thinner, I cut the next

*The nail nick, after grinding.*

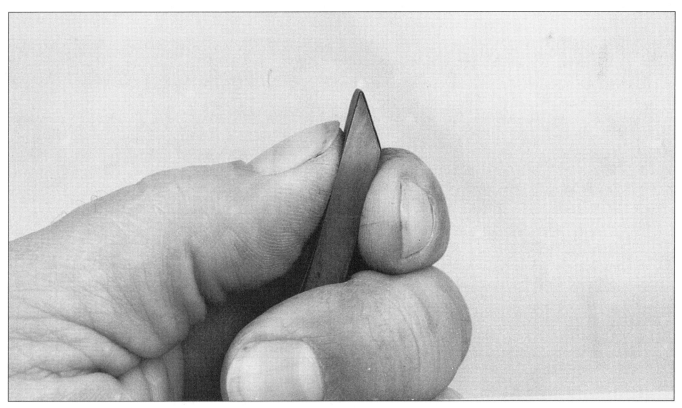

*To prepare for sanding the inside of the nail nick, wrap a slip of 1000 grit paper around a tapered Plexiglas™ block.*

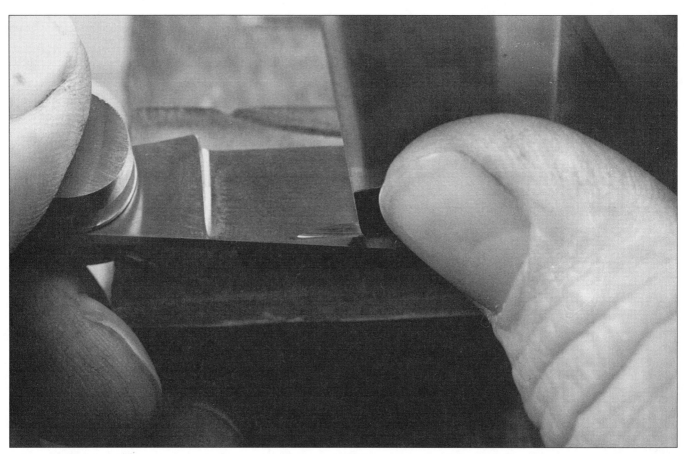

*Use the Plexiglas™ block wrapped with the 1000 grit paper to hand-sand the inside of the nick.*

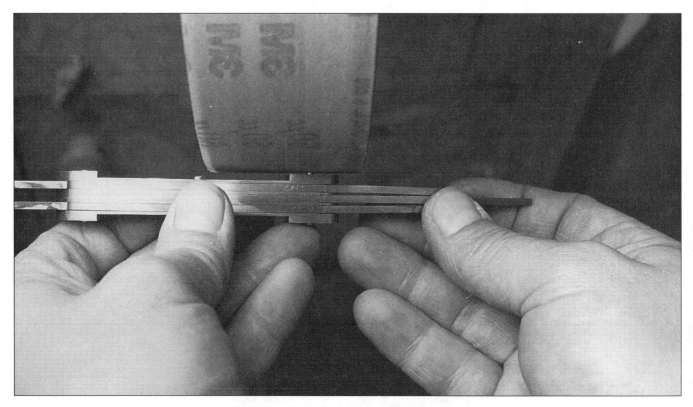

*Preparing for disassembly, grind off the domed heads of the pins.*

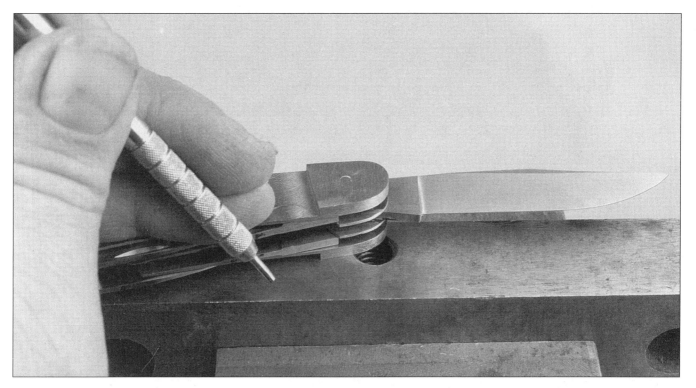

*After grinding off the head, position the pin over a hole in your anvil.*

*Extract the pin by driving it out of the hole with a punch and hammer.*

*Grinding the choil using a custom-made grinder that features a rest with a miter groove cut in it. The miter slide is fitted with a rubber-tipped clamp to hold the blade.*

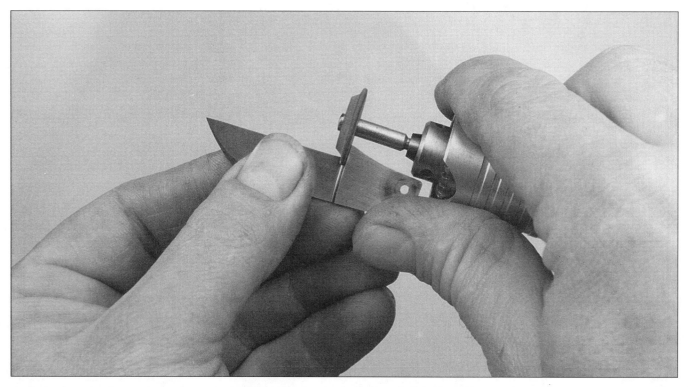

*Polish the choils with a cratex wheel shaped to fit the ground-out area.*

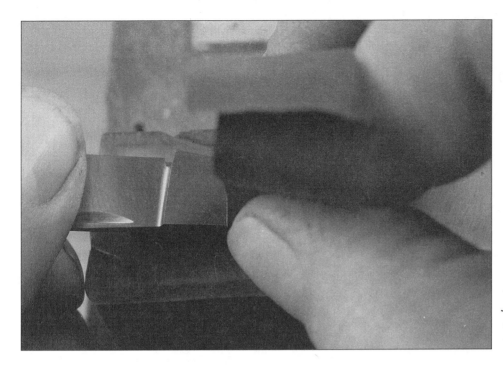

*Hand-sand the blade tang flats with 800-1200 grit Klingspor paper wrapped around a Plexiglas™ block.*

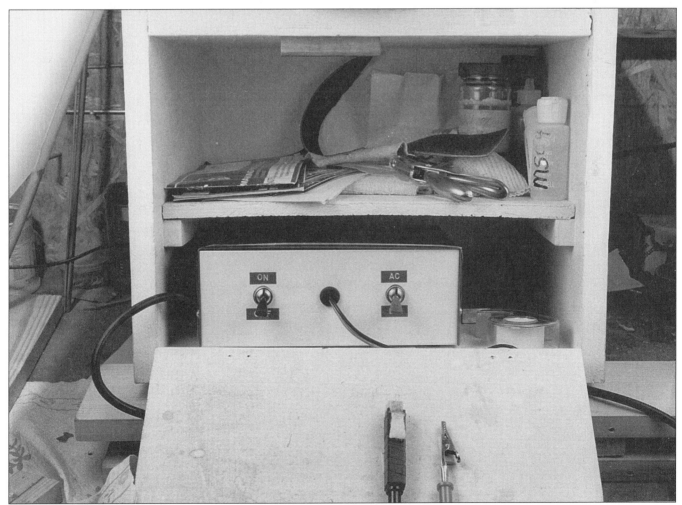

*The etching machine supplies a charge for etching the lettering to the blade. Note that the switch is turned to DC.*

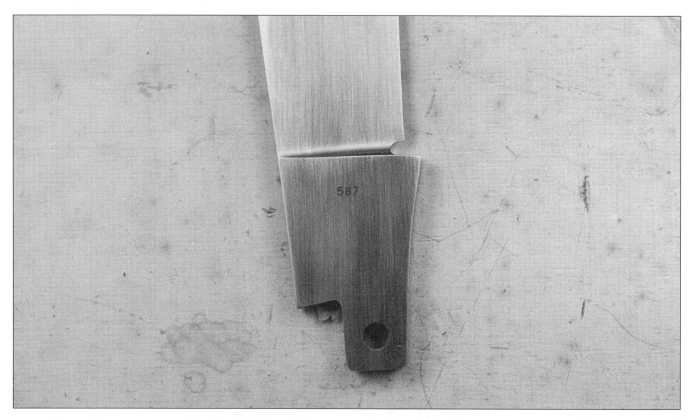

*The project number etched on the master blade.*

*The name after the etching process.*

*Grind the bolsters to the proper shape using an 80 grit Norton Hogger belt.*

*The next step in preparing the frames for the handle material is to drill the handle pin holes.*

serial number from the stencil sheet. The numbers on the sheet are very small and very close together on the sheet. To facilitate handling it and to keep it from sliding around on the surface of the tang, I use a small piece of Scotch™ tape. I first make a hole in the center of the tape with a small paper punch and then position the hole over the number on the stencil. Now I can locate the number where I want it on the tang and attach the tape. I still need to protect the blade from unintentional etching, so I place more tape on any exposed areas around the stencil. With the ground clip attached to the tip of the blade, I apply the wick to the end of the handpiece and moisten it with electrolyte. Making sure that the AC/DC switch is on DC, I turn on the power. For just a couple of seconds, I touch the wick to the stencil. After a few seconds, I repeat the action for another few seconds. The etching should be of sufficient depth at this point. Switching from DC to AC, I repeat the entire procedure of touching the wick to the stencil. Then I turn off the power and remove the ground clip.

I remove the stencil and swab the etched area with neutralizer to remove any remaining electrolyte residue. Now that I've completed etching the number, I want to repeat the entire process with my name stencil. Before we leave the subject of etching and serial numbers, I would point out that I have received very positive response from customers regarding the use of serial numbers.

After drying and oiling the blades, I rub the blade tangs with a piece of fresh 1,200-grit Klingspor that's backed with a rubber block, thus straightening out the 1,200-grit marks.

I'm ready to prepare the frames for handle material. First, I drill the handle pin holes, using a 1/16-inch drill bit, locating them .300 inch behind each bolster, splitting the difference between top and bottom. Any burrs left from drilling are easily removed with a small file.

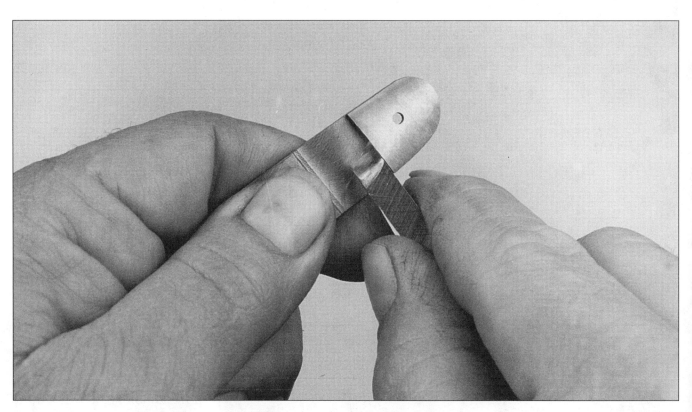

*Deburr the inside of the bolsters with a sharp square lathe cutter.*

*Grind the ends of the handle material—in this case jigged bone—with the Bader belt grinder.*

Now the bolsters need to be ground to the proper contour. I use an 80-grit Norton Hogger belt to shape them, and then I regrind them on a 60-micron belt. All this grinding leaves a burr on the handle side of the bolsters. I use a sharp 1/4-inch square lathe cutter, laying the cutter against the bolster and pushing the cutter forward, trimming off the burr.

Turning to the handle material, I need to thin the bone to the proper thickness. The ends should be around .085 inch, while the center of the material can be thicker. When working with pearl, I want thickness of 1/8 inch to allow for enough contour. If the material starts out too thin, the center of the handle will look flat and unappealing. If this happens, the bolsters must be ground thinner so the pearl can be properly shaped. Alternatively, the thin pearl could be replaced with thicker material.

I cut the handle material to fit against the bolsters, trimming it with a bandsaw for bone, or a Dremel® cutoff-disc if the handle is pearl. Turning to the Bader, and with the power on, I carefully grind the ends to fit *fairly close*, then I shut off the machine and turn the belt by hand for "fine tuning" the fit.

I clean up the handle areas in the frames with lacquer thinner and rub down the blade side of the frame, using a piece of 220-grit paper, which has been glued to the sanding plate. This removes any hole burrs left the last time I disassembled the folder.

I have a piece of bar stock that I've ground flat that I use for gluing the frames

*Fine-tune the handle material to fit fairly close by turning the belt slowly by hand.*

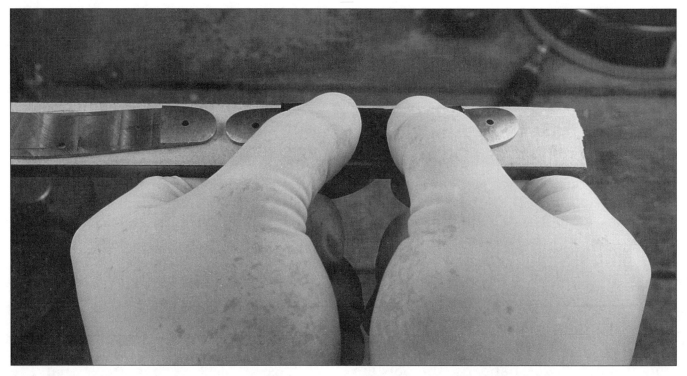

*Glue the handle material onto the frame. The gluing base is flat-ground bar stock covered with masking tape.*

*Apply LocTite Depend® adhesive to the back of the jigged bone pieces. There is no reason to rush—an activator is needed to set the glue.*

when I use bone. The bar has a fresh piece of masking tape put on it that's the width of the bar. This prevents the glued frame from sticking to the bar. The frame is laid on the taped bar, and the handle material is glued onto the folder frame.

I have used LocTite Depend® adhesive for the past seven years and have been very satisfied with its performance. The glue sets up in a couple of minutes, and it creates a truly tenacious bond. It is a two-part product, adhesive and activator. I always apply the adhesive to the handle material and the activator to the frame. After affixing the handle to the frame, I wiggle it around a little to mix the two ingredients; then I hold the parts together for a couple of minutes until they bond. I repeat the process with the other handle and frame. When both are completed, I clean up the excess glue with a small chisel and acetone.

Next, the bone must be drilled for the handle pins. I stand my micro lathe on its end, using it much the same way as a drill press. I position an angle-iron plate on the carriage, holding the frame under the angle. I drill the hole from the inside frame-side through to the outside of the bone. The drilling must be done slowly and carefully or the bone may chip; a very sharp bit is essential. After drilling the front and rear handle pin holes to .063 inch, I run a .064-inch reamer through them to make the tiny domed pins easier to insert. I drill the center pin holes to .093 inch and then ream them out to .0945 inch—again for easy pin installation. The drilling and reaming creates burrs, which I remove with a small file.

Knifemaking with Eugene Shadley

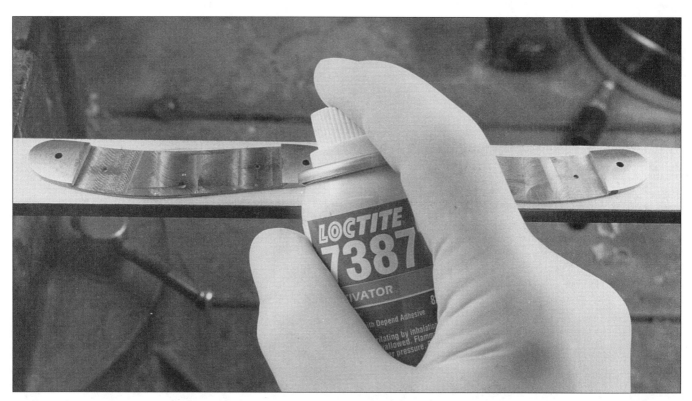

*Spray the LocTite Depend® activator onto the frames. Attach the handle material with the adhesive applied; the bond will be permanent in just a couple of minutes.*

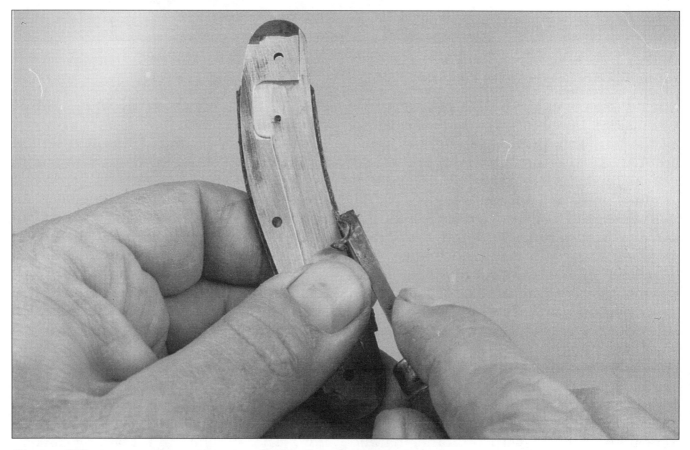

*Clean off the excess glue with a small chisel and acetone.*

*Drill the jigged bone for the handle pins. Use a micro lathe set on end, operating it much like a small drill press.*

*Drill the hole through the bone from the inside of the frame, drilling very slowly with a sharp bit to avoid chipping the bone.*

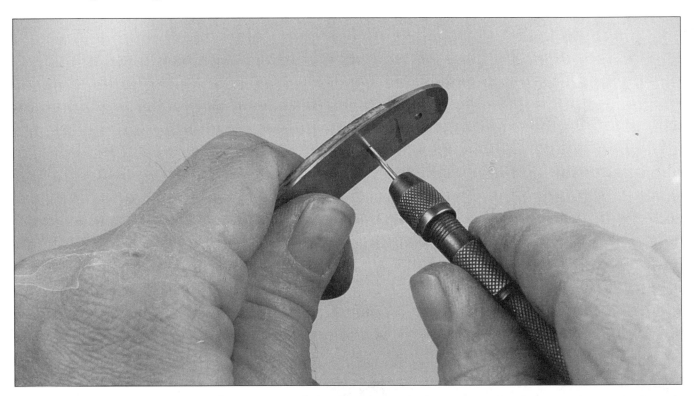

*After drilling the holes to .063", run a .064" reamer through them to make the pins easier to insert.*

*The tip of the 1/16" pin stock is given its domed shape using a head spinner fitted in a 1/4" drill.*

*The domed pin head after shaping with the head spinner.*

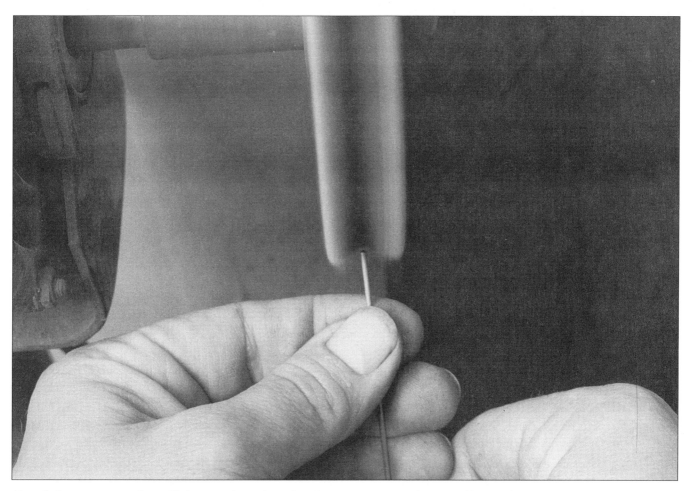

*Sand the spin marks off the pin head with 600 grit paper, then buff to polish.*

*Countersink the pin holes with a Dremel® 1/4" No. 125 cone-shaped cutter.*

To make the handle pins, I place 1/16-inch pin stock in a drill-press vise and a head spinner in a 1/4-inch drill. The head spinner is one that I shaped and Tony Bose heat-treated. Leaving the pin stock about 1/16 inch above the vise and applying a drop of 3-In-1® oil to the head spinner, I start the drill while applying pressure to the top of the pin. A little practice can produce a nice head on the pin. With a small rubber block and a piece of 600-grit paper to remove the spin marks, I sand the dome of the pin and then carefully polish the dome on a sewn wheel, using green-chrome compound.

When I've cut the pins to length, they're ready to be installed, but not before I countersink the bone slightly to prevent the heads from being ground off when I finish the knife. I do this using a pin vise and a Dremel® 1/4-inch cone-shaped cutter #125. Inserting the cutter in the hole, I turn it by hand three times, or enough to make the proper depth countersink on the bone. Turning the frame over, I countersink the 416 with one rotation of the pin vise. This countersinking of the inside of the frame allows a place for the peened material to fill in and form a head on the other end of the pin. I put a nail set that fits the pin head in my large vise. The pin is inserted through the bone side, with the pin head aligned to rest on top of the nail set. I trim the pin to almost flush and gently peen the cut end to form the head on the frame side of the pin. Using a rubber aluminum-oxide wheel, I carefully flush up the frame side of the pin and rub it with a piece of 320-grit paper, backed by a Plexiglas™ block.

After trimming the sides of the handle material almost to the frame, I need to grind, contour and polish the belly. First I stack the spacers between the frames, inserting pins in the pivot holes at the ends and grinding and polishing the belly of all the parts simultaneously. Once I countersink the pivot pin holes, I'm ready

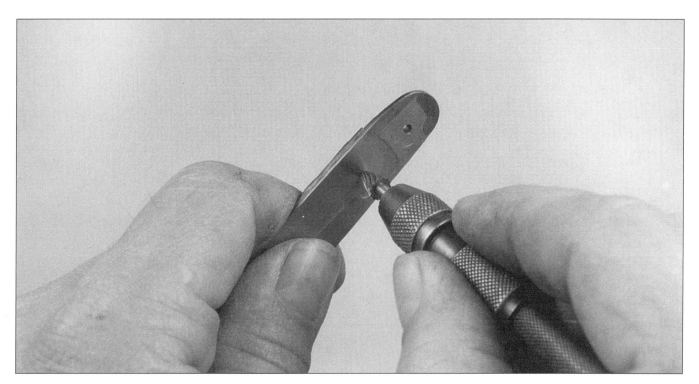

*Countersink the inside of the frame pin hole with one turn of the cone-shaped cutter to create a space for the peened material to fill in.*

*Position one of the frame holes over a nail set secured in a vise. Insert the pin until it rests on top of the nail set.*

*Trim the pin almost to flush and gently peen the cut end to form a head on the frame side of the pin.*

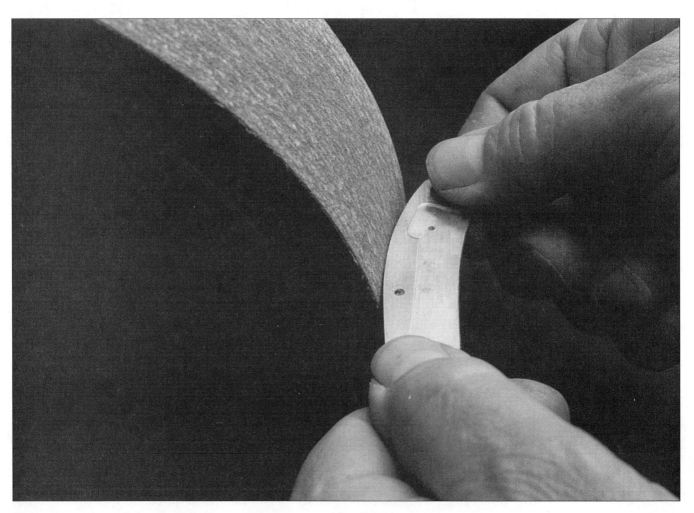

*Trim the sides of the handle material almost to the frame.*

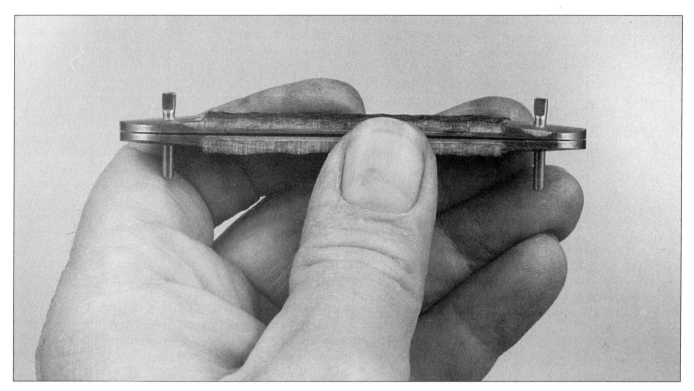

*Place the pins in the pivot holes at the ends of the frames*

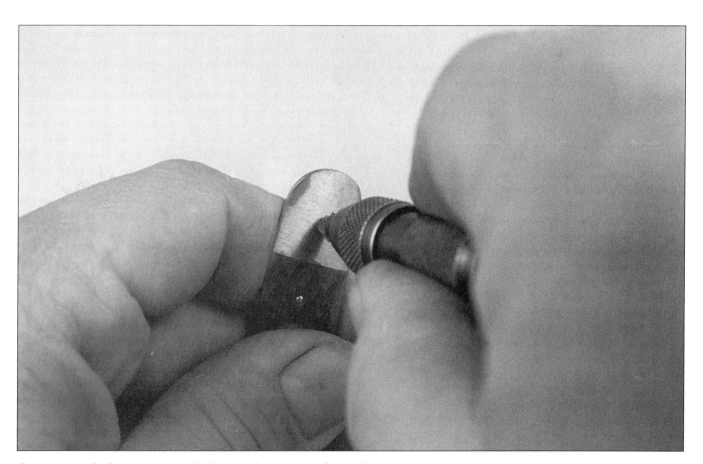

*Countersink the pivot pin holes with a cone-shaped cutter.*

# Knifemaking with Eugene Shadley

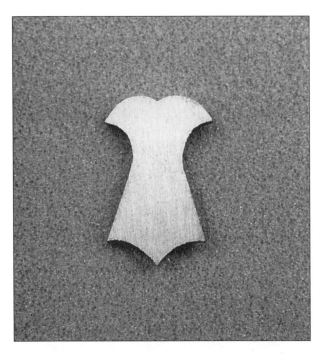

*The sowbelly shield.*

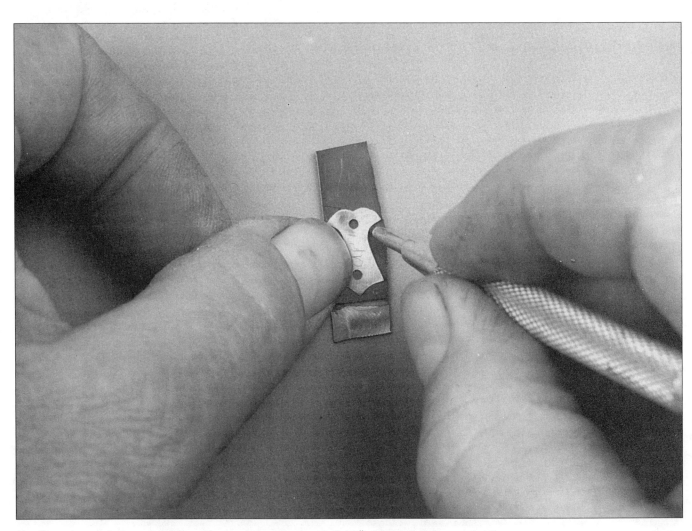

*Scribe around the shield pattern on a piece of .040" 410 stainless steel.*

*Super Glue™ the cut-out and ground shield to the position it will occupy on the bone handle.*

to begin the process of installing the shield in the left-hand frame.

My sowbelly shield pattern is fetched from its hiding place. I place it on a piece of .040-inch 410 and scribe around it. After cutting out the part with the band saw, I move to the Bader and grind as close to the line as I can. At this point, I must remove the ridge that has formed on the down-side of the profile. I simply rub it on the sanding plate, still covered with the ever-present 220-grit paper. Next, I clamp the shield in the drill press vise and file to the line, trying to keep it as symmetrical as possible. I clean it up by rubbing it again on the sanding plate. Placing a drop of Super Glue™ on one side of the shield, I position it on the bone. After the glue sets up, I scribe around the shield, using a sharp needle in a pin vise. The closer I can keep the line to the shield, the better chance I have of getting a real close fit.

Then I detach it from the bone by inserting a tiny eyeglass screwdriver under the shield, popping it off. I put a mark on the bottom side of the shield (as in the sod-layer's reminder: "green side up"). It seems that no matter how carefully I make the parts, they normally cannot be flipped over and still fit.

Using a moto tool and a dental cutter, I remove the material inside the scribed outline. This procedure requires great care and concentration—a slip very likely will result in replacing the bone. Once I've removed the bulk of the material, I switch to a small engraving tool, scraping out to the line. With luck, the shield will fit on the first try. However, luck is generally in short supply in my shop, so I often have to try it several times, removing more material each time, until the shield fits properly. The cavity is then covered with a piece of masking tape, which keeps it

*Scribe around the shield using a sharp needle held in a pin vise.*

*Remove the excess bone from inside the scribed line to create an inlay for the shield.*

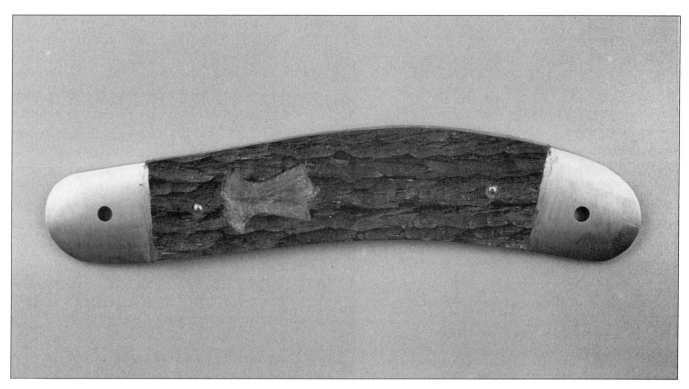

*The completed cavity should fit the shield as closely as possible.*

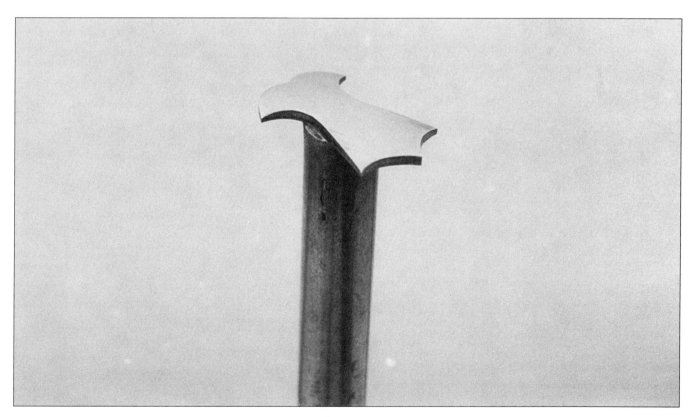

*The shield is glued to a 1/4" rod, then ground and hand-sanded to 2000 grit, creating a slight dome shape on one side.*

clean until I'm ready to install the shield. I prefer the shield to stick up above the bone just enough to allow it to be slightly domed when finished.

To dome the shield, I glue it to a piece of 1/4-inch rod for shaping. First, I grind it on the Bader and then hand-sand it to 2,000 grit, doming one side. Then I heat the rod with a mini torch until I can remove the shield by pushing it off the end. After the shield cools, I clean it by soaking it in acetone. I make sure it is cold before placing it in the acetone—I never feel adventurous enough to risk a shop fire. Personally, I'm okay with letting the insurance company make a dollar.

## When the Handle is Pearl

With pearl, my process is somewhat different. After the pearl is test fit to the frame, the folder is assembled for the final time. All the parts are checked and cleaned, and all the exterior frame pin holes are countersunk and deburred. The assembly block is then put back in the vise. The pins are put in the holes in the block, and all the parts are positioned in their proper order. Once the folder is together, the pins are trimmed. I put shim stock in the pivot ends and peen. This will be covered in more detail when we return to the bone-handled folder (final assembly).

The pin heads are ground flush with the top of the bolsters. This leaves a slight burr on the inside of the handle cavity, which is removed with a lathe cutter. Next, I peen the center pin and file the excess pin flush. Now the frame can be cleaned and the pearl glued on, one side at a time. I apply the adhesive to the

*After the pearl is glued to the frame, grind the pearl flush to the belly and back of the folder.*

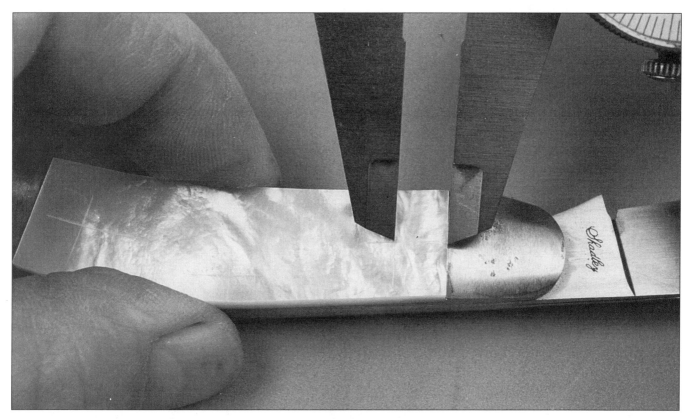

*Using a dial caliper, locate the pin holes .300" from the ends of the pearl.*

*Set the pearl on top of the angle iron rest on the micro lathe and drill the pin holes.*

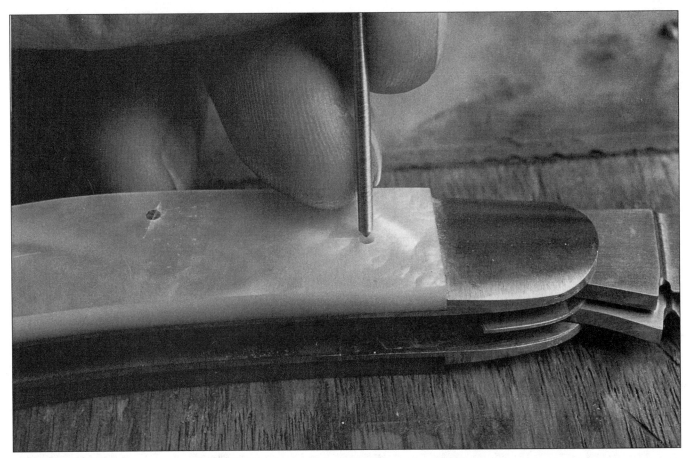

*Glue the dummy pins, made from 1/16" 416 pin stock, into the pearl pin holes. Dummy pins minimize the risk of pearl cracking under the stress of use.*

pearl and the activator to the frame, and then press the parts together, holding them for several minutes. The excess glue is trimmed off with a small chisel; any residue is cleaned up with acetone. I repeat the process for the other side. After the glue has completely set, the pearl is ground flush to the back and belly of the folder.

I leave the pearl flat for now—to make drilling the hole easier. Using my dial caliper, I locate the pins .300 inch from the ends of the pearl, splitting the difference between the top and bottom. The center pin is located in the middle of the pearl from end to end and .200 inch from the top of the back. With the pearl left flat, it can be set on top of the angle-iron rest on the micro lathe and drilled in the normal manner. I've just started using 1.6mm carbide circuit board bits from MSC. They have a short flute length and are solid carbide. This combination helps stop the bit from skating around on the pearl—a real problem, since you can't center-punch pearl. These bits are working well for me now, but I will keep my eyes and ears open; when I find something that works better, I won't hesitate to change methods once again.

Once the holes are drilled, I take 1/16-inch 416 pin stock and bevel the edge to a blunt point. I put Super Glue™ in the pearl hole and insert the pin stock to the bottom of the hole, making sure it bottoms out. Then I cut off the pin—flush—and wipe the excess glue. The process is repeated for each hole.

I use dummy pins in pearl, thereby minimizing the danger of the pearl cracking under the stress of use. Pinning through the pearl, especially the center

When grinding the sides of the pearl, start on a wheel then switch to a slack belt, like this one, going up to 9 micron.

When mounting a shield on pearl, contour the shield on a rounded jig like this, shaping it with blows from a rubber mallet.

pin, could cause the pearl to crack if the folder is snapped shut or the springs are too strong. I know that LocTite Depend® will hold the pearl in place until the cows come home, and I don't have to worry about replacing pearl due to pin stress-cracks. There may be purists among my fellow knifemakers who would argue in favor of pinning completely through the pearl, but they probably have more hair to lose than I can afford. I'll continue to use my method, so long as I'm satisfied with how it looks and lasts.

I have found it best to grind pearl with a new belt, and I never use a belt that's more coarse than 120 grit. When I get close to where the pearl meets the liner, I switch to a 220 belt, since the larger grit belt seems to leave little chips along the pearl-liner joint. This shows up as an unsightly cloudy white line. To prevent this, I use a 220 belt, running on a smooth rubber wheel at the slowest speed on the Bader. The back and belly are both done completely on a wheel. In contouring the sides, I first grind them to 220 on a wheel, then I change to slack belt, going up to 9 micron. I then hand-sand the bolsters and pearl to 2,000 grit and polish with pink no-scratch compound.

Normally, I don't put shields on pearl-handled knives, but when I've done so, I've surface-mounted them. I drill two holes in the pearl through the shield, contour the pearl to final shape, and then form the shield to fit that contour. The shield is placed on a jig that I made just for this purpose and whacked with a rubber mallet to form it to shape. Once the shield is contoured, domed and polished, I spin the heads on two pins and polish the domes. Then I attach the shield and pins with LocTite Depend®. A well-done shield really adds to the appearance of the finished piece. I think it should look like it set down roots and just grew there.

## *Final Assembly*

Now that we've explored the peculiarities of working with pearl, let's return to the project at hand. At this point, the handle has been glued and pinned, and the pivot pin holes have been countersunk. The handle material on the belly of the folder has been ground flush and polished, and the shield and cavity have been prepared. The blades and springs must be cleaned and any tang scratches removed. The folder must be reassembled and once all of the pieces are in their proper places, the pins are trimmed with side-cutters and the ends flattened on the grinder. I make sure to use the air hose to blow out any grit that may have accumulated in the countersink. I leave about 1/32 to 1/16 inch sticking up on either side of the bolsters and about 1/32 inch above the bone on the center pin. Not a precise measurement—I pretty much eyeball it.

After the pivot pins are ground flat, I place .002-inch shim stock on one side of one of the blades. Having placed a shim in front of the pivot pin and one behind, I take the folder to the anvil, along with my pin hammer. I use a very small claw hammer, the face of which I've ground flat and polished. I take the folder pin that I want to peen and place it with one end flat on the anvil and the other end facing up. Taking my hammer (and holding my breath), I lightly peen around the edge of the pin, first on one pin end and then turn it over and do the other side. Now that the heads are formed, I tap on them a little harder. I want to make sure I fill the countersinks on each side. It takes quite a lot of practice to get the countersinks filled so that the pins don't show. There is only one chance to get it right the first time. Any do-overs involve starting again with a new pin. Translation: You get to take the whole thing apart for a purely cosmetic correction.

When the shim stock is removed, the blades will have a little play in them when

*Reassemble all the knife parts, then trim the pins with side cutters.*

*After grinding the cut pins flat, peen with a small claw hammer like this. Note that the hammer face has been ground flat and polished.*

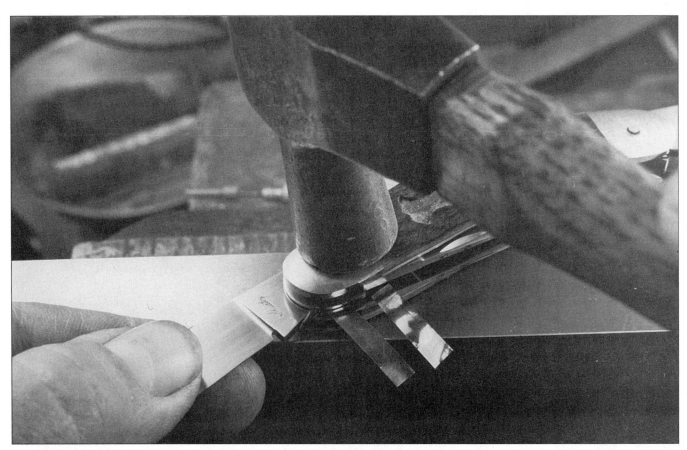

*With the knife held on an anvil, gently peen the pins, working first to form the heads on each side, then tapping harder to fill the countersinks on each side.*

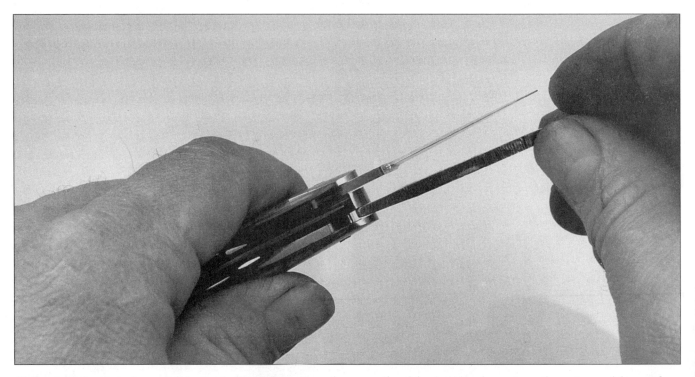

*If the blades bind up after installing the pins, loosen the blades by levering the assembly with a wedge like this.*

*The head spinner, inserted in a power drill chuck, used to spin the heads on 3/32" pins.*

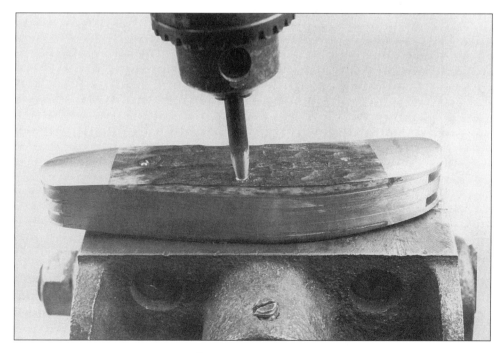

*Spin the center pin with the knife resting on a vise, working on one side, then the other, until both ends of the pin are nicely domed. Of course, you need to hold onto the folder.*

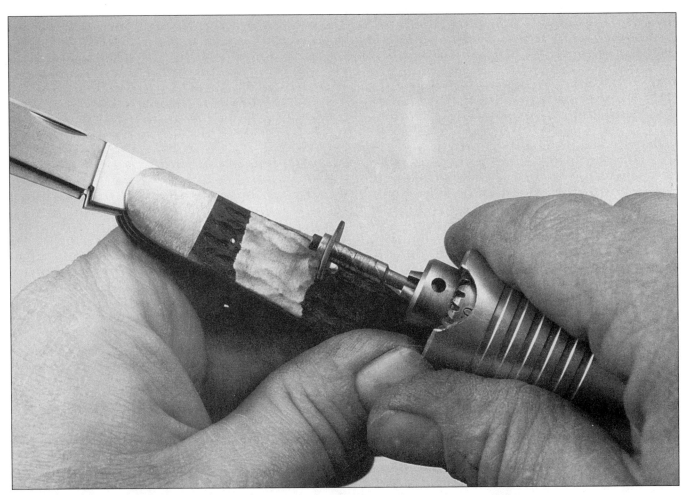

*Clean up scratches on the domed pin heads with a small cratex polishing wheel.*

*Dress down the back of the knife by grinding with the blades on the master end fully open, and the pen and sheep blades halfway open.*

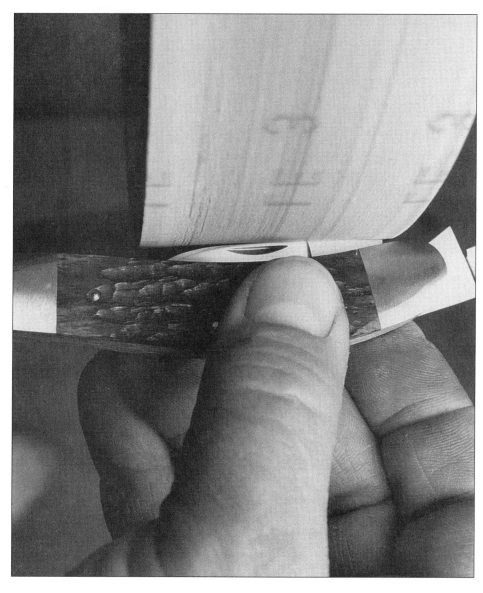

*If you grind too much off the top of the springs you will have to regrind the top of the blades to fit. Here, the top of the pen blade is reground.*

wiggled from side to side. This is taken up by tapping on the pins a little and wiggling to test, repeating if necessary. When the blades are sufficiently tight, **STOP!** An extra tap at this point can bind things up—not a pleasant phenomenon, trust me. When it happens to me (and it does happen to me), I grind the pin heads flush with the bolsters, and loosen the blades with a wedge made for the purpose. With one blade closed, I insert the wedge into the front. I apply a little side pressure and lever the wedge a little, being careful not to bend the spacers. I remove the wedge and try to close and open the blades. Sometimes they work smoothly; if not, I adjust them a bit more. In the worst case, I have to tear the whole thing apart and put in new pins. If this becomes necessary, I center-punch one side of the pin. Then I drill the 3/32-inch pin with a 1/16-inch bit. Now the countersink shoulder has room to collapse and the pin can be driven out.

But let's assume the best case—that I have successfully peened the first end-pin. I repeat the procedure on the other end, and now it's time to spin the center pin. For this, I use another head spinner that I made for 3/32-inch pins. Spinning the center pin is more challenging than spinning a pin in a vise, since now we're working on a pin that's sticking up through a piece of bone. Bone can be brittle and will chip or crack if han-

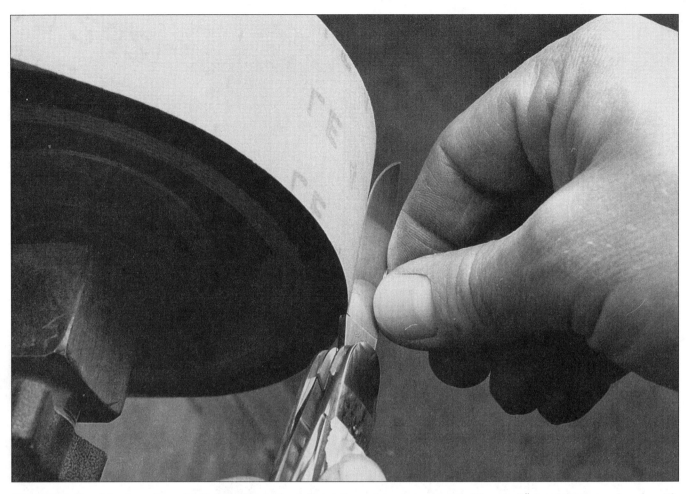

Sharpen the blades on a belt grinder running at the slowest speed. Use a 6" (or larger) wheel with a smooth, squared face.

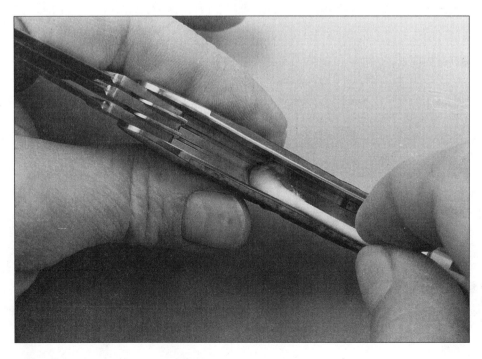

After sharpening, thoroughly clean the knife. Here, the inside of the knife is carefully wiped with cotton swabs.

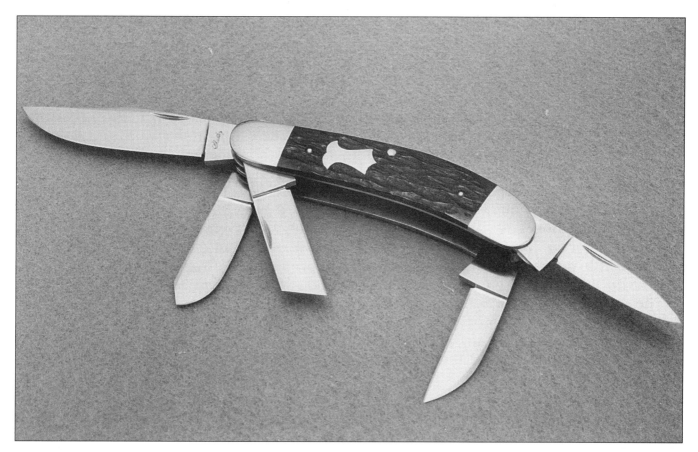

*The finished five-blade sowbelly, No. 587.*

dled carelessly. I know that I must go slow and take it easy.

Laying the folder flat on the vise, pin-side down, I place the oiled end of the head spinner on the pin and start the drill. I spin a little on one side and turn it over and do the same to the other side. I go back and forth until the pins are nicely domed. Then I use a small rubber aluminum-oxide polishing wheel to clean up the scratches and buff it with green-chrome polish on a small Dremel™ polishing-bob.

The bolster pins are now ground flush (if I didn't have to do this earlier to loosen the blades). The back of the folder can now be dressed down flush. To do this, I open all the blades on the master blade end, leaving the other blades closed. With other folders, all of the blades are left open, but this method is necessary with the sowbelly because of the way the pen and sheep blades come out of the end of the folder-body when open. Special care must be taken when grinding on this end to avoid taking too much off the top of the springs. A good measure of caution in this step will save a lot of grief. A mistake means that the top of the blades must be ground to fit again. Once I've got it right, I start grinding, going through the grit sizes 60-9 micron. After 9 micron, I either hand-sand with 1,500-2,000-grit paper or polish on a sewn wheel with pink compound.

The bolsters need to be finished. I go through the grit sizes 60-9 micron and start hand-sanding with 1,500-grit paper. I go to 2,000- and, sometimes, 2,500-grit paper. The masking tape comes off the shield cavity and the shield is glued in, with (of course) LocTite Depend®. The excess glue is cleaned up with acetone. The bolsters are ready to be polished along with the bone.

The blades will have small scratches in them from handling during assembly and dressing down the folder. I remove them by opening the blades and hand-sanding each one with a piece of 1,200-grit paper and a rubber block.

The folder is now ready for sharpening. I use my Bader at the slowest speed and a 6-inch smooth wheel. Larger wheels can be used—remember that the face of the wheel should be in good shape. If the edges and face of the wheel are not square, it will cause problems in sharpening. Rounded edges allow the belt to flex and hinder a proper edge at the choil area. I run the machine in reverse for safety; I'm not partial to puncture wounds, even if my doctor does make house calls!

The sharpening starts with a 60-micron belt followed by a 9-micron belt. I grind one side of each blade, adjust the belt over to the edge of the wheel, and grind the other side. There will be a slight burr left on the edge, which I remove by stropping the edge side to side on a piece of leather.

Then I take the folder to my office for cleaning and final inspection, before shipping or placing in inventory. To clean the folder, I open up all the blades and clean out the inside with cotton swabs and then blow it out with an air compressor. I oil the inside tang/spring joint, and, one by one, I close the blades to half-stop and wipe down the exposed tang flat, closing each blade before moving on to the next. I then clean the inside ends and oil the tang/spring joint. I apply lemon oil to the bone with a swab and allow it to soak in before wiping down the entire exterior of the folder with a clean cotton cloth. Once cleaned, the folder goes into in a small zip-lock storage bag.

With nearly every knife that I sell, I furnish some type of protective covering.

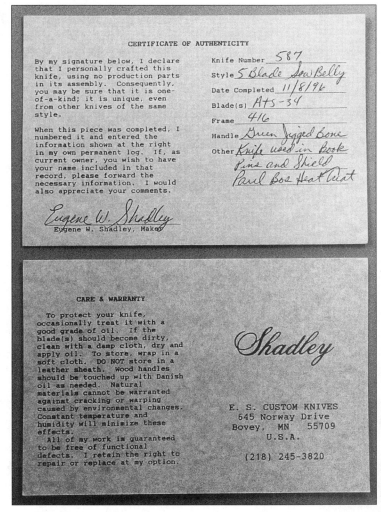

*The certificate of authenticity for Shadley knife No. 587.*

For my fixed blades, I make leather sheaths; but I prefer to use soft cloth slip-covers for the folders, which I used to make myself. While my leather skills are pretty good, my talent with the sewing machine leaves much to be desired. Cindy Sage, a friend with far more sewing skill than I possess, now makes all of my slip covers.

## The Paperwork

The job isn't done until I finish the paperwork. I enter all of the pertinent data in my logbook, complete the Certificate of Authenticity, and prepare the bill and shipping slip if the knife is to be shipped.

*The knife and the book are finished, but my job isn't done until I thank the individuals whose contributions made it possible. To my wife, Bev Schwartz, who translated my hand-scribbled notes into readable text, I owe a huge debt of gratitude (as well as a new arm chair—that was the deal!) Thanks are due also to Terry Davis, my friend and partner in this enterprise, for his willingness to translate his notable skill in knifemaking into words of guidance for the rest of us. Finally, I thank Tony Bose, not only for writing the foreword, but also for his friendship and encouragement.*

# Part 2
# Knifemaking with Terry Davis

# A Great Experience

When Gene Shadley called, asking if I wanted to be a part of this book, my first thought was that it would be a great experience. Now, as the writing nears completion, I feel the same way. I will admit, at the beginning of the writing some time ago, it seemed to be the same sort of great experience as a "root canal," but I feel that this book is needed and I'm glad to be a part of it.

Multi-blade custom knives are just starting to gain their rightful popularity. It wasn't that many years ago when multi-blades were a curiosity at a knife show. I look forward to a time when they will have the same acceptance as other handmades. I think this book will help.

Knifemaking shouldn't have any secrets. The free exchange of ideas and methods is vital to its growth. This book is an attempt, by two multi-blade makers, to shed a little light on some old secrets and on a skill that is being reborn. I hope you enjoy it.

## *That First Knife!*

Why am I a knifemaker? Sometimes I'm not even sure myself, but I'd like to give you some background on myself that may show why it was bound to happen.

I live with my wife, Cheyleen, and daughter, Sara, outside the old mining town of Sumpter in rural Eastern Oregon. It's a timber and ranching area I grew to love while working for the U.S. Forest Service in the 1960s. I grew up in Michigan, but after coming out west the first time, I found it hard to live anywhere else.

Growing up in a family business of excavating contracting provided me with an early exposure to machinery and shop work of all kinds. Years later, my various small businesses of welding, blacksmithing, gunsmithing, and commercial refrigeration taught me a lot of hard-won skills I've used in my knifemaking.

I'm fascinated by tools of all types, from bulldozers to screwdrivers, but my special interest has always been in the old tools, like knives, that have evolved through the years with elegance and simplicity.

I made simple knives during the 1960s and 1970s when I was building muzzleloaders. Then in 1984 a friend, Bill Woodward, gave me a copy of *Blade Magazine*. Nothing had ever caught my attention quite like what I saw there. I wore that magazine out. It opened a world that I never knew existed. I was hooked.

When I began making knives, I tried to copy all the well-known makers' work from the pictures in the magazine. It took quite a while for me to realize that if my designs didn't come from inside me, they just wouldn't work. They sure didn't. I wasn't happy with anything I made that first year.

Then one day in a local hardware store I caught myself staring at a display board of pocketknives. Some kind of folder has always been in my pocket since I was a child, but I had never seriously thought of making one, let alone a multi-blade. I came out of the store that day with a little three-blade stockman knife in my hand and started my first multi-blade that afternoon.

Making that first pocketknife was a real educational experience. The funny thing is that the more knives I make and the more I learn, the more there is to learn. I used to be naïve enough to think the curve would level off. The truth is, if a person is trying to make quality knives, the learning curve just keeps getting steeper.

I actually sold knives at my first show, which is proof that a low enough price will sell anything. I basically paid for my gas. Much more importantly, I came home with enough encouragement and ideas from other makers to keep me fired up for months. A lot of the methods I now use on my knives have come from other makers. Many of the really good tips have come from Ron Lake. I can never thank Ron enough for his ideas and especially for his challenge to make a 100 percent (quality) knife. It's an impossible goal, but it's a good one.

I've been a full-time maker for about eight years. I joined the Knifemakers Guild in 1994 and was made a voting member in 1996.

# Tools, My Favorite Subject

All, the knifemakers I've known have their own unique knife-making methods. They use different tools in their own special sequence. The only constant is the end product, a good knife. Anyone learning knifemaking should let their own creativity show them the method and tools to use. There is no best way; this book contains suggestions only.

For me there is almost as much pleasure in making and modifying my own tools as there is in making knives. Knifemaking is a low-tech enterprise that lends itself to shopmade tools of all kinds. I'd like to show you a few of my tools to give you some ideas on how to equip a knife shop without a lot of expense.

The **belt grinder** is the most useful tool in any knifemaker's shop. I made mine to fit my needs when I first started knifemaking, and then modified it over the years as I saw ways to improve it. It's an old friend that has been through a lot of surgery.

Because I flat grind all my blades, a large flat platen is my grinder's main feature. It's made from a 2-inch-wide bar of hardened A-2 steel with a row of bearings to ease the belt around the angle at the top of the platen. There are sections of slack belt above and below the platen for finishing bolsters and other slack belt jobs.

A vertical grinder layout is necessary because I always stand when grinding blades. This also saves a lot of space in the shop; my shop has none to spare. Unseen in the photo is the grinder base made from a Caterpillar© tractor rear sprocket. These make an excellent base

*This belt grinder was shopbuilt to fit Terry's requirements.*

for all sorts of tools. I use them whenever I can get them.

The grinder uses 2-inch x 72-inch belts, has a 10-inch serrated contact wheel, and develops a belt speed of 3200 surface feet per minute. The 1-1/2-horsepower motor was obtained in trade for one of my first knives. I'm sure that I got the better end of that deal.

The **small radius grinder** was probably the most difficult of all my tools to make because of belt tracking problems. It has a 1-inch and a 3-inch contact wheel, and it

*The small radius grinder has 1" and 3" contact wheels.*

*A different kind of surface grinder, which uses a belt rather than the usual grinding wheel.*

took quite a while to get both of them tracking right. The 2-inch x 72-inch belts run at a slow speed of 1100 surface feet per minute to help prevent heat buildup.

Like all my shopbuilt tools, it is built to fit the way I like to work. I always use the grinder standing up and grind the underside of springs and blade tangs freehand. The handiest thing about this grinder is that it is always ready to work and doesn't have to be changed over to a different contact wheel for every job.

A **surface grinder** is not an absolute necessity to make folders, but it sure isn't much fun without one. Mine is made from an old turret lathe, which provides the base and the ways. The upper assembly is shopbuilt and allows the solid metal contact wheel to be raised and lowered to grind a part to the desired thickness.

During use, the magnetic chuck holds the blades as it is moved back and forth under a belt running on the 8-inch contact wheel. This contact wheel started life as a regular rubber covered wheel, then lost its rubber. It was installed and then trued, while rotating, with a cutting tool held on the magnetic chuck. I found by experience that a belt speed of about 1200 surface feet per minute gave the best finish and didn't cause excessive heat buildup. This grinder does not need coolant like a regular surface grinder.

Another difference from a conventional surface grinder is that I don't have to worry about this grinder injuring me. When grinding wheels explode on a surface grinder the result can be ugly. Belts occasionally break on belt grinders—which is startling—but the damage is only to the knifemaker's nerves.

The **disc grinder** is an indispensable tool. Bob Lum demonstrated that to me long before I knew how handy the disc grinder could be. Mine is shopbuilt with two 9-inch discs that turn at about 850 RPM. The slower-than-usual speed helps control heat buildup. (Has anyone noticed that the words "heat buildup" keep being repeated?) Abrasive paper is attached with 3M® spray adhesive, which stays tacky, allowing paper changes. This is handy

*Terry calls this disc grinder the handiest tool in his shop.*

because this tool goes through a lot of sandpaper. After the sheet is stuck on the disc, a knife is used to trim the paper around the circumference of the disc.

The feature I like most on this tool is its ability to swing the work rests out of the way for changing paper. Free access to the disc allows me to use it for flattening handle scales. This feature makes it a pleasure to use.

My **heat treating furnace** is a Paragon KM-14 that I rebuilt with a digital controller. It is much easier to work with now and doesn't have to be watched minute by minute as it did before. I still use a Fluke digital thermometer hooked to a separate thermocouple in the heating chamber. This digital thermometer is accurate to 2°F over the entire heat range and is a valuable monitor in addition to the digital controller.

The condition of the heating coils must be watched carefully in these furnaces. After the coils have spent some time at the upper end of the furnace's heat range, 2000°F, they lose their ability to heat the furnace as quickly as before and must be replaced.

My **sub-zero freezer** was built using my experience in commercial refrigeration. Living in a rural area has prevented me from having access to liquid nitrogen or other cryogenic liquids. They say necessity is the mother of invention. In this case necessity led to building my own freezer.

The freezer uses two 1-horsepower, low-temperature-rated compressors running in series. This generates an almost total vacuum in the evaporator (cold chamber). It will reach -180°F in about fifteen minutes and will hold that temperature for the three hours required in the heat treatment cycle. The freezer has been amazingly trouble-free for the last eight years and it allows me to cold treat a batch of blades and springs for about 25 cents worth of electricity.

There is no picture of this beauty. If you are wondering what it looks like, just take a look under your refrigerator—that's it!

The **Bridgeport® Milling Machine**, while not a knifemaking necessity, allows me to make a better quality knife. Features like the integral liner/bolsters and the cut nail nicks I use would not be possible without a milling machine. I highly recommend some sort of mill to anyone starting to make folders. Even a small mill will allow the making of tools and the use of knife features that would be impossible without it.

It is important to remember that this mill, like all my equipment, is hand controlled. My hands and experience guide it. If I lose my concentration, it is all too happy to make junk parts.

My mill required considerable rebuilding before it could be used accurately. With the aid of a good friend, Tom Buffenbarger, I learned the old skill of machine scraping. Using this technique, I leveled

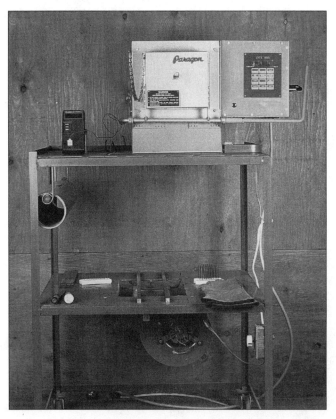

*The Paragon KM-14 heat-treating furnace, fitted with a blower underneath for air quenching.*

*The Bridgeport milling machine makes it possible to use knife features like integral liner/bolsters, that would be nearly impossible to do otherwise.*

*The felt wheels on the buffer are expensive, but provide excellent control. Note the dust collection system.*

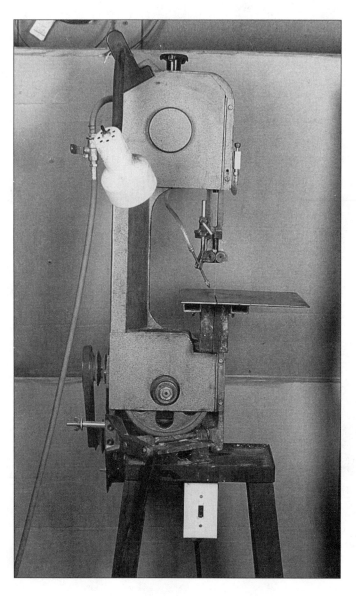

*A shopbuilt stand allows the bandsaw to be used standing up.*

*The spring press allows a slip joint knife to be assembled easily.*

*A good steady—a small, cutler's anvil—makes working on pocket knives a pleasure.*

and trued the ways both on the knee and the table axis, making the mill almost as good as new. This technique is a valuable skill. It allows the repair of otherwise unserviceable tools at a bargain price.

The **dust collection system** is not a tool, but I wouldn't want to do without it. The use of my lungs is something I've never taken for granted. Of the many dangers of knifemaking, dust is the most insidious—don't make knives without protecting your lungs!

My system uses a Dayton 2C864 radial blade blower with a 1/2-horsepower motor. It is visible in the lower left corner of the belt grinder picture. This baby will pull the doors of the shop open when it's turned on! There is also a wall fan behind the belt grinder that works especially well in my long narrow knifeshop.

The **buffers** in my shop get much less use than they did a few years ago, because I now hand-finish the majority of each knife, but there will always be a few jobs left for the buffer. I use felt wheels almost exclusively now, because of the better control they give. The 10-inch wheels I use are expensive, but worth every penny.

My **band saw** is of some forgotten brand. It worked faithfully for years in my welding shop before I started knifemaking. Now it sits on a shopbuilt pedestal which allows me to use it standing up. I use 18 tooth per inch bimetal blades from Suffolk Machinery for 1/8-inch and thicker stock. For cutting all the thinner material I use 24 tooth per inch Simonds bimetal blade stock that I bought years ago on sale at MSC in a 250 foot roll. I use this to make up blades with silver soldered splices. Making up my own blades saves about 50 percent of the cost.

A good **spring press** lets me put knives with stiff springs together easily. The one I have is made from a No. 366 Pana-Vise with a brass "finger" added to push the springs into position.

The **steady** is a small, cutler's anvil. I bought mine about ten years ago from R.E. Roberts, the maker. It is hardened, surface ground, and has every feature a knifemaker could want. I would recommend it to anyone.

There are other miscellaneous tools around the shop that I have accumulated over the years. I will describe these later when we make the knife.

The important thing to remember is that none of the tools I've described are absolutely necessary to make knives. Specialized tools just make the job easier and more enjoyable. The tools used for my first folders were nothing but a belt grinder (built with an old 2 x 6 and a washing machine motor) and a drill press. I admit, I would hate to go back to basics now, but the point is, it would still be possible to make knives.

# The Project: A Wharncliffe Whittler

## Design and Layout

Because this book is about making multi-blades, you might wonder how I arrived at a decision on what to make. That's easy! I love to make knives that I would carry and use. Let's look at what would comprise my ideal multi-blade pocketknife.

I prefer small knives, 3 to 3-1/4 inches long when closed, and slim in width so they are easy to carry. I've always liked sunken joints and round handle ends. They give a knife a very pleasant feel and save pants' pockets. My hands tell me a lot about what I like in a knife. If that magical feeling is there, I know the knife is right.

Straight-edge blades always get the most use in the pocketknives I carry. Hunting or other types of use would require another blade shape, but for my everyday use—whittling, wire stripping, newspaper clipping, opening boxes, and all the rest—the straight-edge blade is by far the most useful.

A knife design that satisfies all my requirements in a very elegant package is the whittler, in particular the Wharncliffe Whittler. It's been one of my favorite designs for years.

Lord Wharncliffe, the presumed designer of the blade shape in 1830s England, must have liked straight-edge blades too. (Maybe he got a lot of splinters; this blade design is the world's best splinter picker!) I personally like both the straight-edge blade and the way the lines flow from the serpentine handle to the point of the blade. The small pen blade and cut-off pen blade complement the whole package and are perfect for all kinds of light cutting tasks.

The type of whittler we'll build in these pages utilizes a tapered divider that separates the two springs. These springs both bear on the tang of the main blade on their forward end, and each of them bears on a small blade tang at the rear.

This whittler has a tapered, classy look. When the blades are closed it will be noticed that because of the taper, the small blades nestle closely to the main blade and there is little wasted space in the knife handle.

The real challenge in this particular Wharncliffe Whittler is that this will be the first one I've ever done. Although I've made whittlers for years, the knife we'll build here is the first in the Wharncliffe style. It's the prototype knife. Talk about staying focused!

Let's start with the handle layout, because everything else has to fit there. Begin with a smooth serpentine handle shape, 3-1/4 inches long, drawn on 1/4-inch grid graph paper. The paper makes it easy to get the length and proportion just right (if it's more convenient, a drawing can be made to a larger size and reduced on a photocopier).

The pivot pin holes should be located halfway from top to bottom on the bolsters for both the main and small blades. The main blade end pivot hole is about 5/16 inch back from the end and the small blade pivot hole is about 1/4 inch back from the small end. The center pin hole,

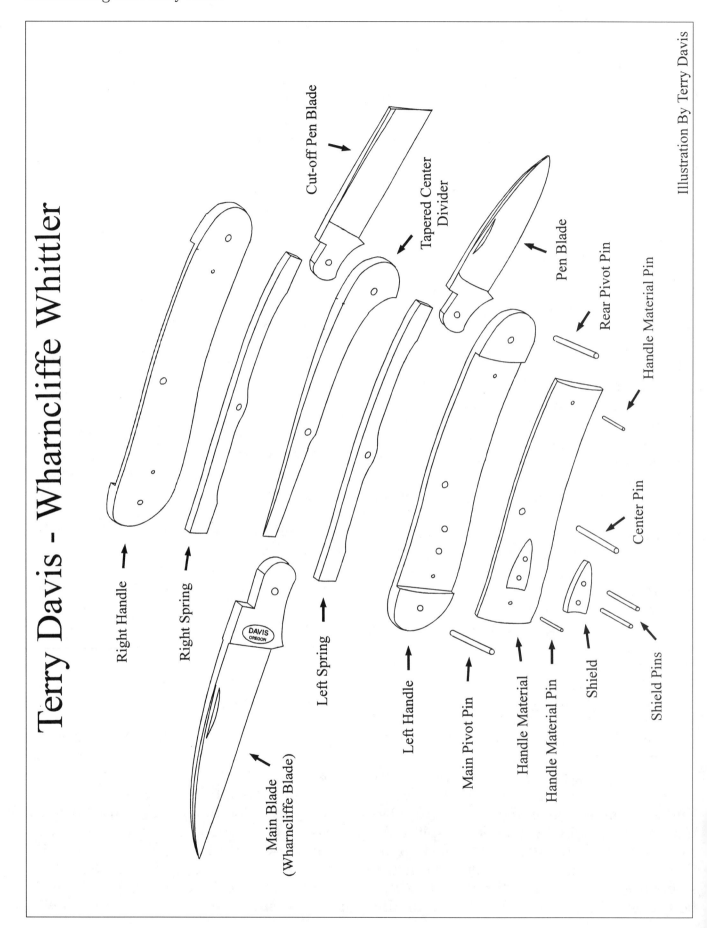

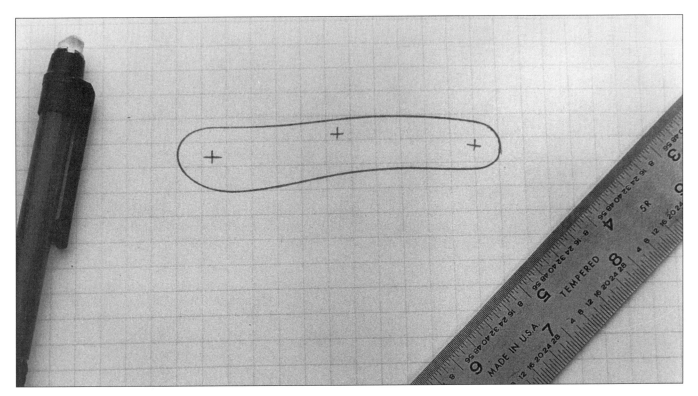

*Sketch out a nice smooth serpentine handle shape.*

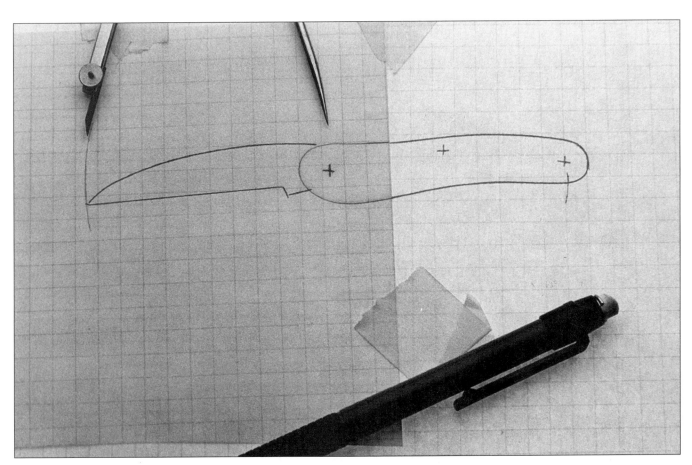

*Swing an arc with the compass to determine the maximum blade length.*

Knifemaking with Terry Davis

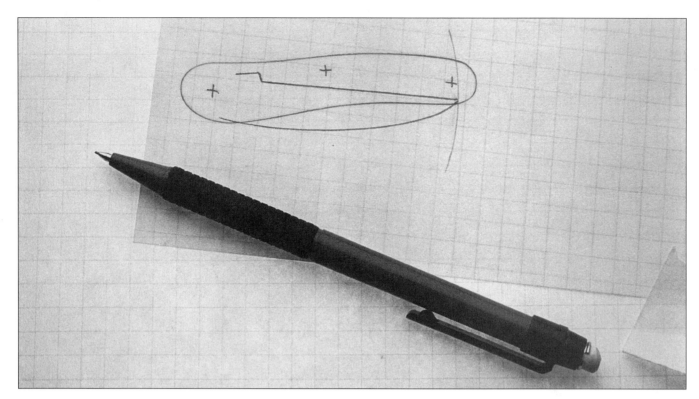

*Swinging the blade drawing to the closed position shows that there will be enough room in the handle for the blade.*

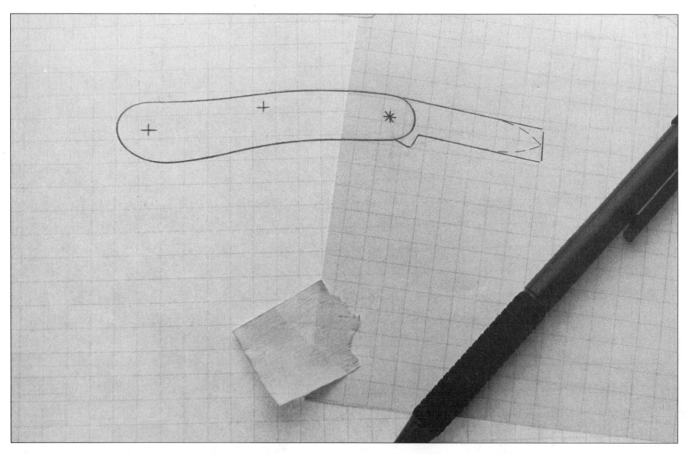

*Draw the small blade in a form that can be modified to several blade shapes.*

as a general rule, is halfway between the front and rear pin holes and about 1/8 inch down from the top line of the handle.

A light table makes the blade layout easier. Tape the handle drawing to the light table. Put a cross mark on another sheet and place the cross over the main pivot hole on the handle drawing. With the point of a compass in the main pivot hole location, swing the compass to the small end of the handle and adjust it to the point where the tip of the blade should fall. Now swing an arc where the tip of the blade will be when open. Draw the bottom line of the blade and then draw the top line curving to a point, starting from well up on the bolster.

Draw the kick in its approximate location (it will be refined later). I always leave enough room for my logo on the tang. Exactly where the kick is located becomes more important on a blade with a lot of weight forward. A long spear or spey blade, especially when there is not much clearance in the handle, requires a kick that's further out on the blade. Because the Wharncliffe main blade is fairly light towards the tip, the kick location is not critical and can be fairly close to the front bolster.

Now swing the blade drawing to the closed position, remembering that the tip will go between the tangs of the smaller blades. It looks like a good fit. Check for room for the spring at the tang end of the blade.

The two small blades are next. Follow the same procedure as for the main blade, except these blades will be drawn in the form of a single blade blank. By doing this, the blank can provide any shape of small blade needed when the knife is being built. A length of 1-5/8 inches from pivot to end is about right for this blade.

## Patterns

Now that the basic layout of the knife is done, it's time to start refining the design and making the actual patterns that will be used to make parts for the knife.

At this point, an explanation of the patterns to be made might clarify the process. They each have a purpose. I don't make them just for fun.

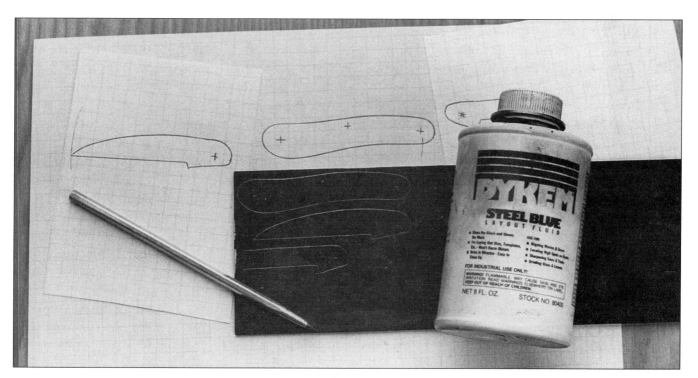

*Transfer the design on paper to the precision ground O-1 steel.*

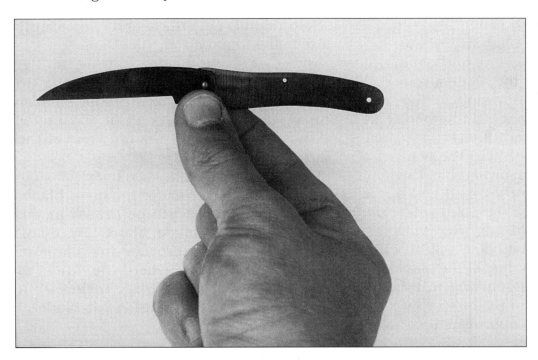

*Hold the patterns up to the light to check that their profiles are smooth.*

The O-1 steel patterns are made directly from the design on paper. These thin, soft metal shapes are easy to work with and give me something I can hold in my hand. I have a hard time designing from paper. My hands are best at telling me what needs to be done. When the O-1 steel patterns are finished they will be used to make the hardened A-2 steel patterns. They will then be set aside to check knife parts against as the knife goes together.

The A-2 steel patterns are the ones that do all the work. They are used to scribe parts for shaping, to drill holes and to set up the milling machine. They are hardened to maintain their accuracy.

There are other specialized patterns and templates that I use to set up to cut nail nicks and shield pockets.

I'm sure that there are those who will wonder, "Why use patterns at all?" When I started making folders I made a lot of knives without any patterns. Then as my favorite cowboy poet says, "My brains came in."

I found that the second knife of a particular design always turns out better than the first. That's because I can see the weak areas on the first knife and do a little fine tuning here and there on the second. If I couldn't reproduce that first design, I couldn't refine it. It's as simple as that. I try to build the best knives I can, and I need all the help I can get. Good patterns really help.

We'll use precision ground O-1 steel 1/16 inch thick for the first outline patterns. Coat the stock with layout fluid. This will give an even blue coating that will show marks clearly. Place the paper patterns over the O-1 steel. Use a sharp pointed scribe and pierce the paper every 1/16 inch or so to leave a mark in the layout fluid coating. Now just connect the dots to outline the patterns. Cut out the pieces with a band saw, leaving a little extra room around the outline. Grind the patterns to shape, being sure not to grind inside the lines. Lightly center punch all pin holes, and drill the appropriate size holes.

For knives of this size, using 416 stainless fittings, I use a #48 drill for pivot pin holes and a #52 drill for center pins. Pins this size in the completed knife are definitely strong enough. Larger pins would only complicate the design by forcing the blade tangs to be larger.

I make all of my setup pins from the shank ends of worn-out drills. I break off

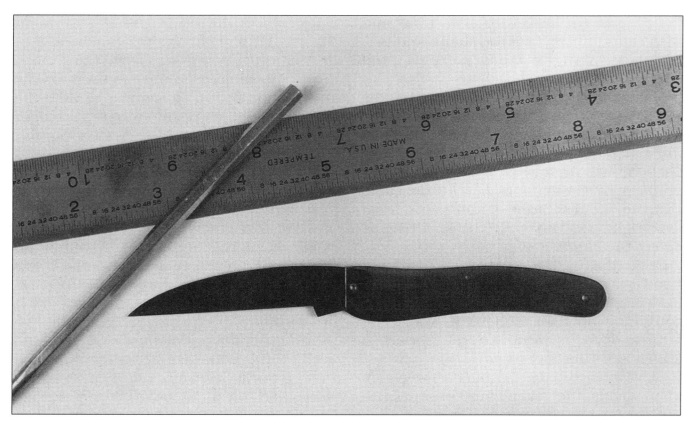

*Scribe the position of the "run up" where the top of the blade and handle meet.*

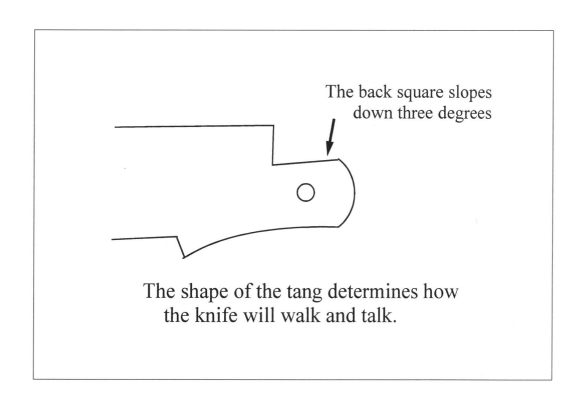

The back square slopes down three degrees

The shape of the tang determines how the knife will walk and talk.

the part of the drill with the spiral and then round the ends of the shank and polish them. These make great pins, and remember, "We recycle."

Pinch the main blade and handle together with thumb and forefinger and, with the blade in its open position, hold the patterns up to the light or a light background. Do they look lumpy or out of proportion? Do the lines flow? If not, take them to the disc grinder and make them look right. Do this with the small blade end of the handle too.

Once the outlines of the blades and handle are looking good, it's time to start designing the tangs of the blades. This will determine how the knife will "walk and talk" and is the most interesting part of knifemaking for me.

Start by laying the handle on the bench and placing the main blade over it with the pivot pin in place. Scribe a line straight down from the intersection of the top of the blade and the front of the handle. This is traditionally called the run up. The bottom line of the "back square" is scribed sloping down at an angle of about 3 degrees to the blade end, a little more than 1/16 inch above the pin hole. File the run up and back square to the scribed lines. The same procedure is used for the small blade blank. Remember to hold the blade in the open position that looks right before it's scribed.

The round end of the tangs are scribed next. Adjust a sharp set of dividers to 0.11 inch. Measure from the rear edge of the pivot hole, scribe a radius on both blade patterns, and then grind to the line. This radius is slightly larger than it will be on the finished blade. When the knife blade blanks are ground to shape this will assure that this important area isn't undersized.

Next we'll make the hardened patterns used to drill holes and scribe parts. Precision ground A-2 steel 1/8 inch thick works very nicely to make these patterns. Layout fluid is used to make the lines show up better in the photos, but I usually don't need to bother with it at this stage because the scribed lines are easy to see.

Start with the O-1 main blade pattern positioned at the top of a bar of A-2 steel.

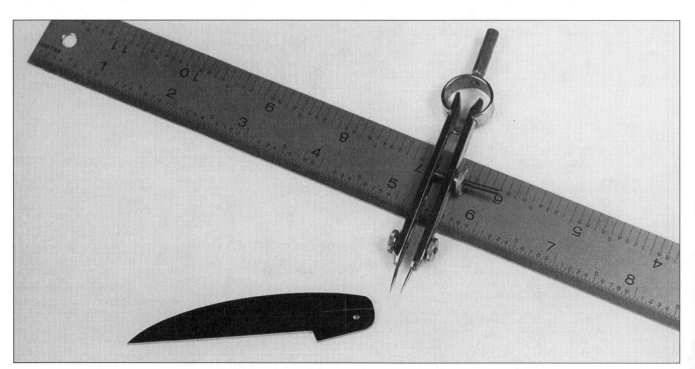

*Scribe the round end of the tang 0.11" from the back of the pivot hole.*

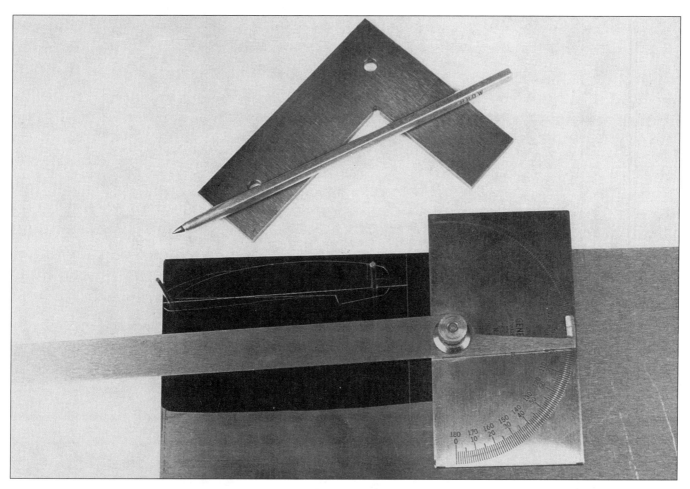

*Locate the holes for the hardened main blade pattern just on top of the scribed line.*

Align the cutting edge parallel with the upper edge of the bar. Scribe around the blade and in the pivot hole. Lightly center punch and drill the pivot hole with a #48 drill. Then scribe a line vertically down behind the tang at right angles to the top of the bar. Using a metal protractor, I scribe a line at a 93 degree angle to the vertical line, going just under the pivot hole and on out past the tip of the scribed blade. At a convenient location outside the outline of the blade, center punch and drill a #52 hole just above the line. Finally I scribe a line from the tip of the blade around the #52 hole and back to the front of the kick. This piece is sawn outside the scribed lines and ground to shape, leaving a little extra in the back square area.

To cut the back square of the main blade to its finished size, place a #48 pin in the pivot hole and a 1/16-inch pin in the hole at the tip. Place the blade, complete with pins, on top of the jaws of the milling machine vise and tighten the vise. Using a woodruff key cutter that has been modified with a 5 degree relief on the cutting face, I cut the back square to the scribed lines, nice and square and clean.

Without a milling machine, this area can always be hand-filed. This, after hardening, makes a template that can be used to accurately file the back square area. I filed this area on blades for years until I discovered that the precision and repeatability of the mill allowed my knives to walk and talk more consistently.

The same exact procedures are used to make the pattern for the small blade. Once they are cleaned up on the disc

*Cut the "back square" in the main blade pattern.*

*Drill the pivot holes in the handle pattern.*

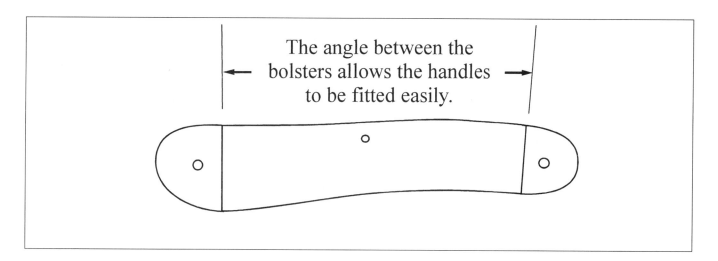

grinder, these blade patterns are set aside to be heat-treated.

The handle pattern is the next order of business. I place the 1/16-inch O-1 steel pattern on the bar of A-2 steel and scribe around it, then cut it out and grind to size. I always leave about 1/32-inch allowance outside the line. Leaving a little extra ensures that a little nick won't make me grind the handle too small. Place the O-1 pattern carefully over the A-2 handle pattern blank and drill the holes to the required sizes. Use the O-1 pattern only once as a drill pattern since it is soft steel.

Next the lines are scribed for the bolster locations. These locations are determined by my old standard, T.L.A.R. (*That Looks About Right!*). With the A-2 handle pattern lying on the bench with the main pivot to the left, I visualize about where the back edge of the main bolster should go, and scribe a heavy line in the steel. Moving to the rear bolster, I scribe another heavy line. These lines will allow setting up the mill to cut the bolsters in the same place on each knife. I like to have the edges of the bolsters at an angle to each other. This makes it easier to get a close fit of the handle material to the bolster.

The holes for the handle material pins are next. I use a 3/64-inch pin on a knife this size, because I feel this size is in proportion on small knives. I turn these pins from 1/8-inch rod with the lathe when using 416 stainless. When using silver or gold the proper size is available as wire. Measure out from the scribed bolster line about 3/16 inch, center the location from top to bottom, and drill the holes with a 3/64-inch drill.

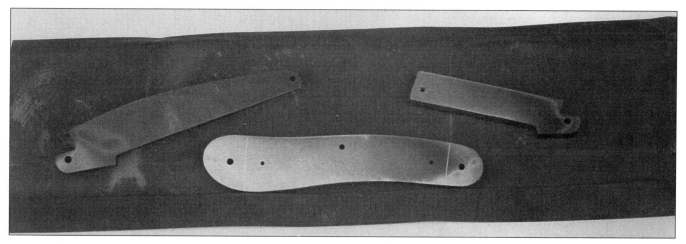

*The hardened patterns are shown fresh from heat-treating.*

With the A-2 steel patterns now complete, they are heat-treated to harden them. The patterns are first sealed in heat-treating foil to protect them from oxidation at high temperature. Then they are loaded into my Paragon furnace for one hour at a temperature of 1750°F. Finally they are air quenched, which cools them quickly to harden them. I always leave them very hard, as quenched, to give the drill holes in the pattern extra durability.

## Starting the Knife

I begin the handles by making the integral liner and bolster units. About five years ago I started making these from a single piece of 416 stainless. It's a superior method for making handles. There are no solder joints, as in my early knives, to cause problems at a later date, and I feel the knives go together a lot tighter and straighter.

I heat-treat the 416 stainless material. As received from Fry Steel, the supplier, 416 bar stock is not heat-treated and will not finish as well or be as rust-resistant as after proper heat-treatment.

Cut two pieces of 1/8-inch thick x 3/4-inch bar stock to 3-3/8 inches long and seal in heat-treating foil.

I use a temperature of 1800°F for forty-five minutes with an air quench and, after cooling completely, temper at 1200°F for two hours.

Surface grind the pieces of 416 on one side. This will be the inside of the completed handle. Now place the hardened handle pattern on the ground side of the piece of bar stock and scribe around it, remembering that a "mark" and a "pile" side are needed. Grind the 416 handle blank to shape. Align the hardened handle pattern carefully back on the ground side of the blank. Clamp and drill the holes to the required sizes using cutting fluid and

*A holder fastened to the drill press keeps drills and reamers organized and handy.*

*Screw machine drills and good cutting fluid will help you produce accurate holes.*

sharp drills. I place a pin of the proper size in the first hole drilled to be sure that the pattern doesn't shift.

This is a good time to talk about drills and drilling. I use 118-degree screw machine drills almost exclusively. These are little short guys. I've found that they don't chatter or wobble nearly as much as the longer standard type of drill. I never drill a hole deeper than 1/4 inch anyway, so drill length is not a concern. Drills with a 118-degree point will drill holes very close to the drill size, whereas the 135 degree drills I've used tend to drill oversize holes.

I buy drills, ten or twelve per package, and throw them away when they start to get dull. I don't pretend to be able to sharpen a drill as well as the factory in these small sizes, and any error in sharpening will almost always result in an oversize hole and a ruined part. I do check new drills with a magnifier, however, since I have found them with obvious defects right out of the package.

By using the short drills, a good cutting fluid like COOL TOOL, and a hardened pattern, I get accurately drilled holes and knives that go together easily with no misalignment.

After that little harangue on drills, it's time to make the jig plate used to mill the liner and bolster units.

I start with a length of 3/4-inch x 4-inch hot rolled mild steel bar about 10 inches long. This is cleaned up and trued with a carbide flycutter in the mill. A piece of equivalent sized precision ground steel could be used, but is expensive. A piece of hard alloy aluminum plate can also be used if available.

With the new jig plate in the mill vise, checked for flatness, drill and tap a line of 10-32 holes the length of the plate, cen-

*Prepare the jig plate to mill the handle sides.*

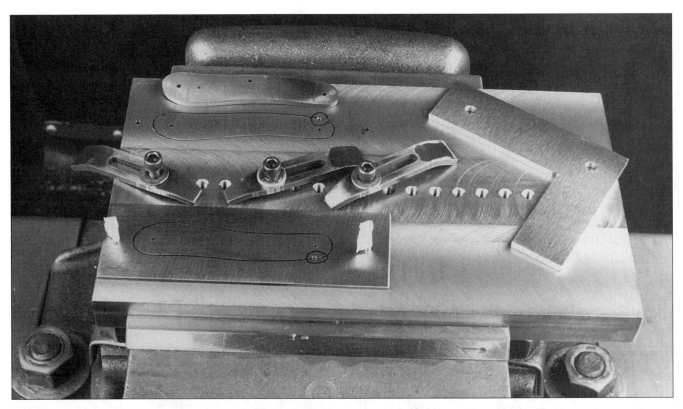

*A template allows the jig plate to be drilled so both handle sides can be milled on the same plate.*

*Milling the mark side handle. Move the center clamp as necessary to hold the handle flat to the plate.*

tered, about 3/4 inches apart. These tapped holes are for the clamps that will hold the liners flat as the mill cuts.

I drill the jig plate so that the handle sides can be milled with the bolsters at an angle to each other. This is done by having a choice of two holes for the rear pivot hole on the jig plate. The lower hole is used to mill the front bolster edge and the upper hole positions the handle side to cut the rear bolster. I use a small template to locate the holes on the opposite side of the jig plate so the pile side handle can be milled utilizing the same setup of the mill.

With the handle blanks profiled and drilled, they can now be completed. First, set the mill to cut to the bolster lines using the scribed lines on the hardened handle pattern as a guide. Then the depth of cut is set. For this knife, the liner portion of the handle will be cut to a thickness of .030 inch.

I use two flute 1/4-inch cutters to mill the handle recess since they don't seem to chip as badly when cutting 416 stainless. I always use a small fine Arkansas stone to put a small radius on the outside of the cutting edges. This spreads the cutting forces and helps the tool last much longer. Good milling cutters are expensive.

With everything set up, cut the recess between the bolsters. I take about a 0.100 inch cut on each milling pass and use a mist cooler which also blows the chips away so they don't dull the cutter. Working with care the milling is completed and a nice liner/bolster unit is ready for handle material. One down!

To mill the pile side handle, use the other set of holes in the jig plate. This will make a mirror image of the mark side handle.

There are several other methods of making the liner/bolster assemblies if a milling machine is not available. They can be soldered together using nickel silver or brass liners and nickel silver bolsters. Nickel silver is a much more forgiving

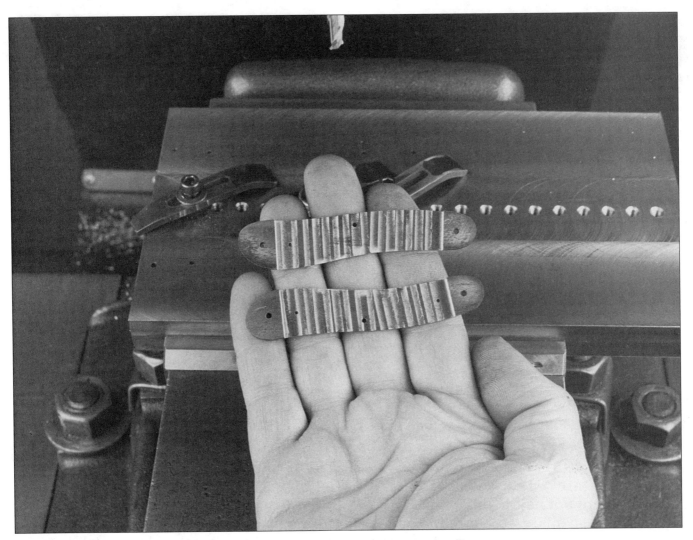

*A matched pair of handle sides ready to be fitted with jigged bone.*

material when it comes to peening pins and finishing bolsters than 416 stainless. I recommend that anyone starting to make folders use nickel silver.

Liners of 410 stainless and bolsters of 416 stainless can also be soldered together. The problem with any soldering job is getting all of the corrosive flux cleaned from the joint. If it's not completely removed it will come out sooner or later and ruin your day.

## *Blades and Springs*

I've used ATS-34 steel for about ten years for the majority of my knives. The fact that it is stainless is important to almost all of my customers. What is important to me is that it makes good knives with excellent edge-holding qualities. I've tried quite a few other stainless steels and found none with the near ideal mix of qualities of ATS-34. Now that 154-CM steel (a near twin to ATS-34) is becoming available in thinner bar sizes, I plan to use it for some trial knives.

I buy my steel from two sources. Sheffield Knifemakers Supply stocks the 3/32-inch hot-rolled sheet ATS-34 that I use for all the small thin blades on my multiblades. The thicker ATS-34 bar stock for whittler and Sunfish main blades comes from Tru-Grit.

Cutting out the blades and springs always makes me feel that I've started a knife. After all this writing, let's make a whittler.

ATS-34 barstock, 3/16-inch thick, is used for the main blade. This thickness allows the blank to clean up to a thickness of .150 inch. Lay the hardened main blade pattern on the bar stock and scribe around it. Then cut out the blank with the band saw and grind to the scribed line. Surface grind the blank to a thickness of .150 inch, being careful to take an equal amount off each side of the rough bar stock.

The small blades and springs are cut from 3/32 inch hot rolled ATS-34 sheet. Place the small blade pattern on the steel and scribe two blanks. Then cut out the pieces using the band saw and grind to the scribed line.

Scribe the two springs by placing the hardened handle pattern on the 3/32 inch steel and scribing around the top of the pattern. After removing the pattern, scribe a line about 1/4 inch under the top line, curving with the top line. Cut out both blanks with the band saw and grind to shape.

Drill the center hole in the springs with a #52 drill after aligning and clamping the hardened handle pattern over each spring. These blanks will be ground as necessary after heat treatment to form the finished springs.

The blade blanks are now drilled and readied for the machining steps. Place the hardened blade patterns over each blank. Then, clamp and drill the pivot hole and the small hole at the tip of each blade blank. Install the modified woodruff cutter in the spindle of the mill and the locating fixture on the mill vise. I use the hardened blade patterns to set up the mill to cut the back square.

The ATS-34 main blade blank with the correct size pins in the holes is placed on the vise jaws and slid to the right until it

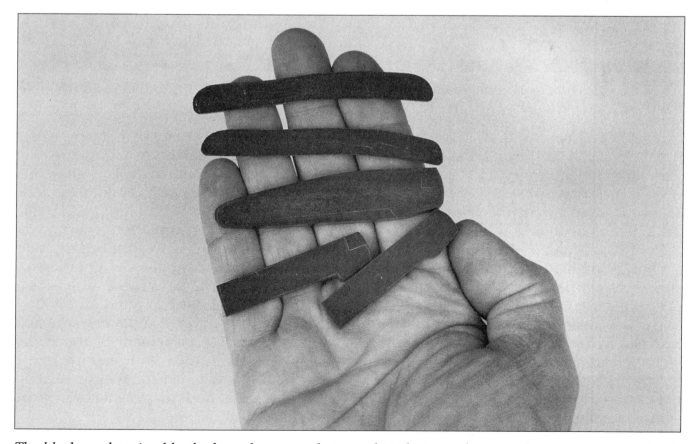

*The blade and spring blanks have been rough-ground to shape and are ready to be made into knife parts.*

*Set up the mill with the pattern to cut the back square on the main blade. The locator rod on the vise acts as a "stop" so that all parts will be identical.*

contacts the locator rod. Tighten the vise. Use a slow feed on the mill and lots of cutting fluid to cut the back square. With care, a nice finish results. The same procedure is used to cut the back square in the small blade blanks.

As an alternate to milling, the back square of the blade blanks can be filed to shape. Each hardened blade pattern is placed on the matching drilled blank and the pins inserted. This assembly is clamped in the vise and the back square is filed; be very careful to keep the filed area square with the hardened pattern. This is not easy on a wide tang like the one on the whittler main blade. Fine sandpaper should be used with a backing block to remove any coarse scratches left by the file. These could act as stress risers when the part is heat treated.

The nail nicks come next. I use a shop-made flycutter with a 3/16-inch square toolbit of Tantung G material. ATS-34, even in unhardened condition, is still very tough material and hard to machine. The Tantung G has the right combination of hardness and toughness required to cut a clean nail nick. I used to grind all my nail nicks but was never completely satisfied with the accurate placement of the nicks on my smaller knives. With the method I now use it's easy to get the nick in exactly the right spot, and they're always clean and sharp.

I use a small jig plate to hold the blade blank in position vertically to cut the nail

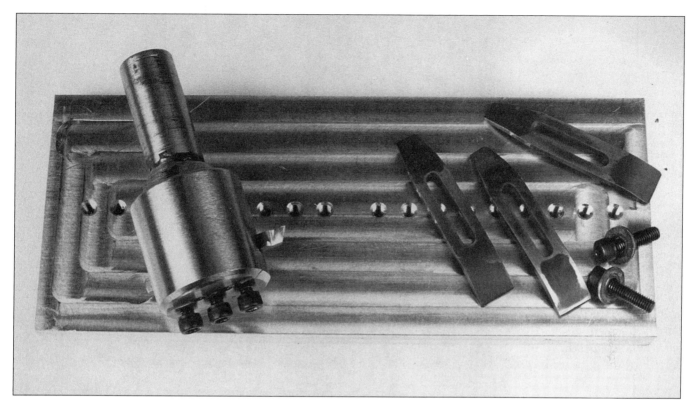

*The flycutter and jig plate are needed to cut the nail nicks.*

nick. I make these jig plates from aluminum that I salvaged from a section of truck frame. This is hard alloy material that machines like butter. A piece of hot rolled mild steel bar about 1/2 inch x 4 inches x 8 inches would work as well but would be much harder to work with.

Prepare the jig plate by drilling and tapping a line of 10-32 holes lengthwise in the center of the plate. These will be used for the clamps that hold the blade blank in place.

A tip I'd like to pass along here is the use of spiral point or "gun" taps. They are primarily used in through-drilled holes because they literally "shoot" the cuttings out the other side of the hole ahead of the tap. I use them with a cordless drill and it takes only seconds to tap a hole. Because they are easy to use, I use lots of tapped holes in my jigs anywhere I might need them.

Using 1/16-inch O-1 stock, make a copy of both of the hardened blade patterns as nail nick placement templates. Then, place the original O-1 blade patterns on each template and scribe their outlines. Doing this lets me see where the finished outline of the blade will be. I mark the location of the nail nicks with a felt-tip pen and scribe a line across the top of the marked nail nick. File a notch from the center of the nick to the top of the template to use when setting up to cut the nail nick.

Lay the nail nick template for the main blade on the top of the jig plate with the nail nick side out, centered between the threaded holes and the top of the plate. Clamp the template in place. Using a #48 drill, drill through the pivot hole and insert a pin. With the template free to move, make the scribed line at the top of the nail nick on the template parallel with the top of the jig plate. Clamp the template in this position. Drill through the small hole at the tip of the template and insert a pin.

Follow the same procedure for the other blades, remembering that with a whittler

Knifemaking with Terry Davis

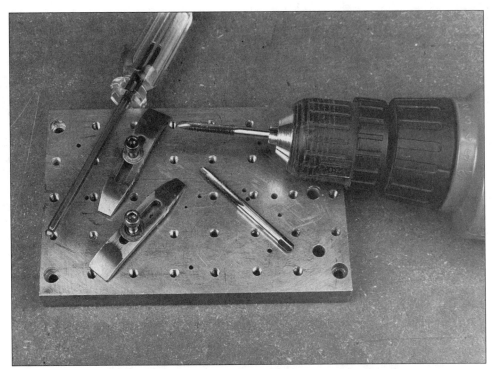

*"Gun" taps make short work of threading holes. The ball point allen driver is the slickest way to tighten the allen screws in jigs.*

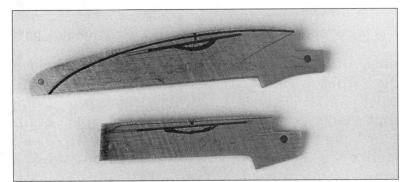

*The nail nick templates allow the nicks to be cut very precisely.*

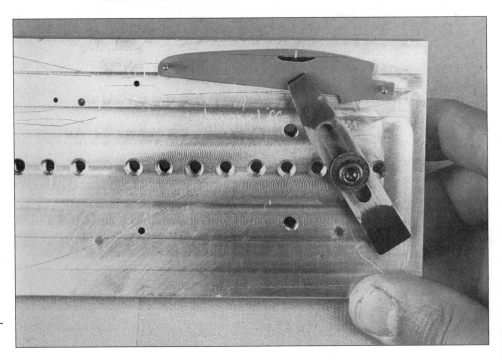

*Position the nail nick template on the jig plate.*

118

*Use the alignment tool to set up the mill to cut the nail nick.*

the nail nicks are on opposite sides of the small blades. Each small blade has to have its own setup on the jig plate.

With the nail nick jig plate done, it's time to cut the nail nicks. Clamp the jig plate in the mill vise. Install the template of the blade that will have its nail nick cut. Align the flycutter with the filed notch in the template to center it side to side on the nick location. The alignment tool I made is illustrated in the photo. The flycutter is then turned so that the toolbit is toward the template. It is raised or lowered so that the top of the toolbit lines up with the scribed line at the top of the nick location. With the mill spindle and table locked, the nail nick can now be cut easily.

With the mill turning at its slowest speed—80 RPM in the case of my mill—bring the turning flycutter carefully forward until it just makes a mark on the blade blank. Zero the mill's dial and, using lots of good cutting fluid, cut the nick. For this knife the main blade nick will be cut .070 inch deep and the small blade's nicks will be cut .040 inch. The main blade nick is cut much deeper because the blade will be ground with a step-down from the thickness of the tang. Once the main blade bevels are completely ground, the nick that remains will be no deeper than on the other blades. Cut the nicks in the small blades in the same fashion, remembering that the nicks are cut on opposite sides of the two small blades. I have quite a collection of blades with the nicks on the wrong side and in the wrong place. Little lapses of concentration can sure cause a lot of extra work.

There are several alternate methods of cutting nail nicks. My early knives had nail nicks cut with a small thin cutoff wheel turned by a Dremel® tool. This made a nick that was rough on thumbnails but simple to accomplish. Nicks can also be cut with a grinding wheel dressed

Knifemaking with Terry Davis

*Cut the nail nick using a slow speed and lots of cutting fluid.*

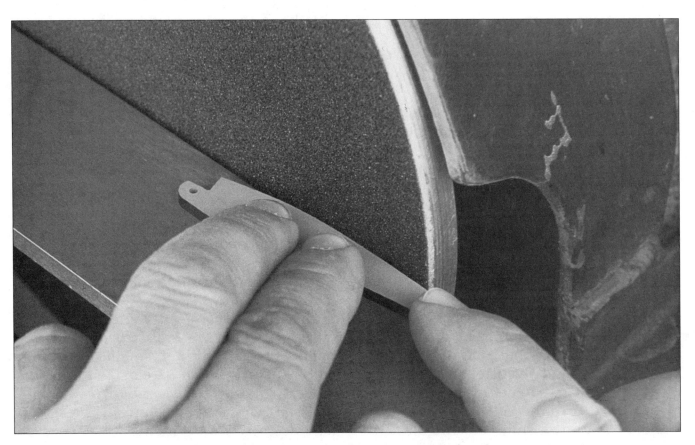

*Finish grind the outlines of the knife parts with 120 grit paper before heat treatment.*

to an edge and turned by a drill press. The blade is clamped in a fixture and slid across the drill press table to cut the nick. Possible ways to cut the nail nicks are limited only by the knifemaker's ingenuity, however, the requirement on a multi-blade knife for the nicks to be in a very precise spot and to a set depth complicates most methods.

I now ream the pivot holes in all the blades with a #46 reamer using lots of cutting fluid. This reamer leaves a hole that is .005 inch larger than the #48 pivot pin. When the pivot pins are set they will swell slightly and provide a nice fit in the reamed holes in the blades.

The blade blanks are now scribed with the O-1 steel pattern and ground to their (almost) finished outline. This removes the hole near the tip that was drilled for use with the various jigs. With a blade like the Wharncliffe master blade, I like to leave the point area a little full and rounded above the edge, otherwise it would be very easy to grind the tip away by accident. The tip area will be brought to a finished point after the blade is completed.

Finish-grind the blanks to their final profile on the disk grinder with 120 grit paper before heat-treatment. There should be no grit scratches left from the earlier profiling steps. Coarse scratches will act as stress risers and can cause cracks when the steel is quenched from the hardening temperature.

I do not do any grinding of the blade bevels before heat-treatment. On small knives it's just as easy to do the grinding after hardening, and it makes the construction of the knife a lot less complicated.

## Heat-Treatment of Blades and Springs

Like a lot of makers, I have strong feelings about the heat-treatment of my knives. I do my own heat-treatment because I want to have control over what I feel is the most important attribute of a knife, its edge holding ability.

There is absolutely nothing wrong with the work of the well-known heat-treaters. I don't want to give the impression that they don't do excellent work, but the fact is, I've always wanted to do everything possible on my knives myself. Because I tell customers

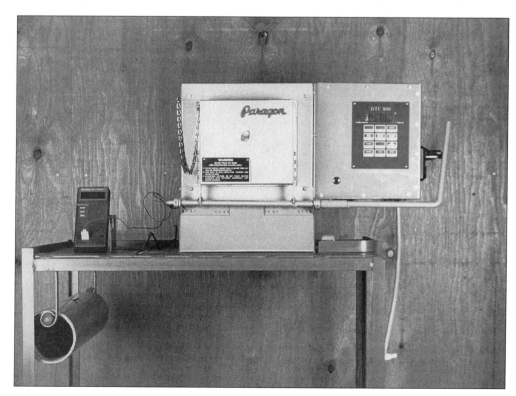

*A good furnace is essential for consistent heat treatment of stainless knife steels.*

that I make the knives, I'd also like to tell them that I take responsibility for them.

The heat-treatment of stainless steel doesn't have to be a complicated task. I feel the minimum requirements are not excessive, but they **are** requirements:

### The Right Equipment

It's almost impossible to do good, consistent heat-treatment on anything but a simple carbon steel without a good furnace and some way to do a cryogenic quench. Accurate temperature control, that is repeatable, is the first priority.

### Time

It takes time and a lot of blades to get the process figured out, and when a different steel is being tried, the learning process starts over. During this learning period it's not ethical to sell the work.

### The Resolve to Throw Out Anything Doubtful

If anything goes wrong with the heat-treating cycle, the steel will not be completely right and should go in the trash.

My own particular terror is the "planned power outage," the kind the local utility tells me about the *next* day in the paper. ATS-34 will be ruined if the heat-treat cycle is interrupted. One heat to critical temperature is all it will stand without a degradation of its qualities.

I make test coupons that I heat-treat with the steel and later break to check the grain structure. What I call a coupon is a small strip of the same metal being heat-treated that has notches filed in the sides so it will break easily in a particular spot. I've found these to be invaluable for quality control. Sometimes what they show is surprising. If the coupon doesn't pass the test, neither does the rest of the steel in the load. Bad day!

With the blades and springs in their finished shape, let's proceed with the heat-treating. The first step is sealing them in heat-treating foil. This foil is made from a high temperature stainless alloy and protects the knife parts from oxidation at high temperatures. It will also cut you like a sheet of paper, only about ten times worse.

Strips are cut from the bulk roll, folded in half lengthwise, and trimmed to even

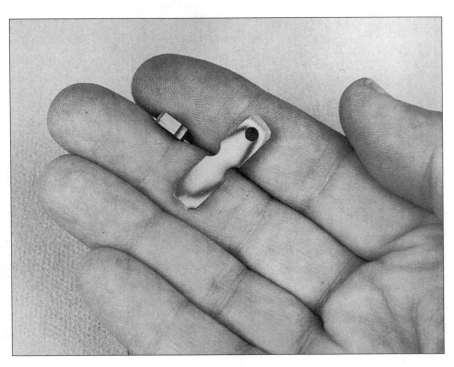

*Test "coupons" allow you to check the steel's grain structure at various times during heat-treatment.*

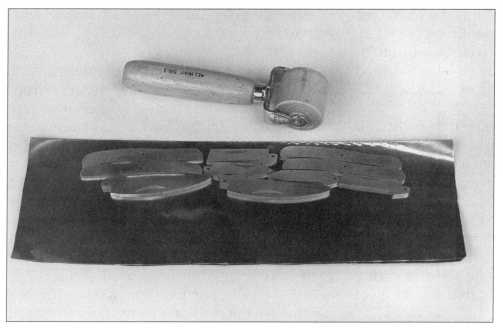

*The knife parts are laid out just as they will be placed in the heat-treating foil. Use the roller to crimp the folded edges of the packet after the parts are enclosed.*

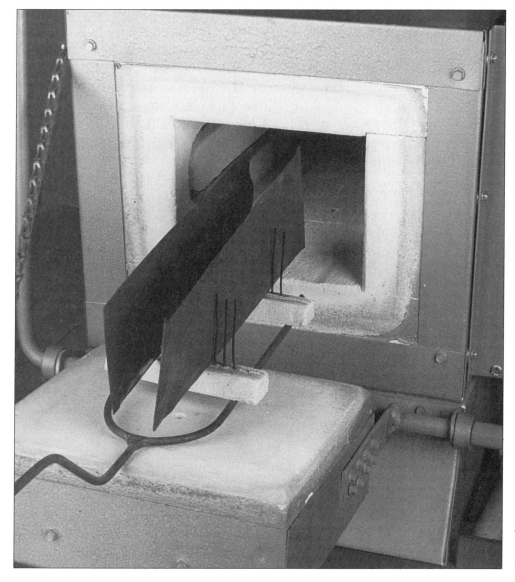

*Place the foil packets in the furnace for heat treatment.*

the edges. The blades and springs are placed in the foil as compactly as possible, one layer thick, for an efficient air quench. The open edges of the foil are then folded over three times and crimped with a small roller.

I load the foil packets on a rack and place them in the furnace with a fork made for the purpose. I try to choose a day for heat-treating that has no interruptions planned. My furnace has a digital controller, but the steel still has to be taken out for the quench and also several times during the tempering cycle. I have to be there in the shop at a given time. Heat-treating days are a good time to catch up on a lot of little jobs.

My heat-treating cycle for ATS-34 has been worked out over the last eight years. It includes a stress relief cycle that I've found to eliminate almost all of the warpage that can occur during the quench.

To stress-relieve ATS-34, I first heat the steel to 1200°F. I let it "soak" for thirty minutes to equalize the temperature. Next, I heat it to 1500°F for fifteen minutes. I then let the steel cool slowly with fifteen minute stops at 1400°F, 1300°F, and 1200°F.

With the stress-relief completed, the hardening cycle begins immediately. From 1200°F, I heat the steel to 1975°F at a rate of 5000°F/hour and hold at that temperature for forty-five minutes. The entire load is then taken from the furnace and air quenched (moved around continuously in a stream of air). I use a squirrelcage blower that is rheostat-controlled to generate the air stream for the quench.

The **quench** is one of the most important parts of the heat-treat cycle. The blade parts in their foil packets come out of the furnace at a yellow heat. A dark green welding visor and heavy leather welding gloves are required equipment when I open the furnace door. It is essential to the hardening process that the steel be quickly cooled below 1000°F. At that temperature the steel in the packets darkens and loses its red glow, and I know I can relax. I've learned through experience how much airflow is required for the right cooling rate. Luckily the ATS-34 knife

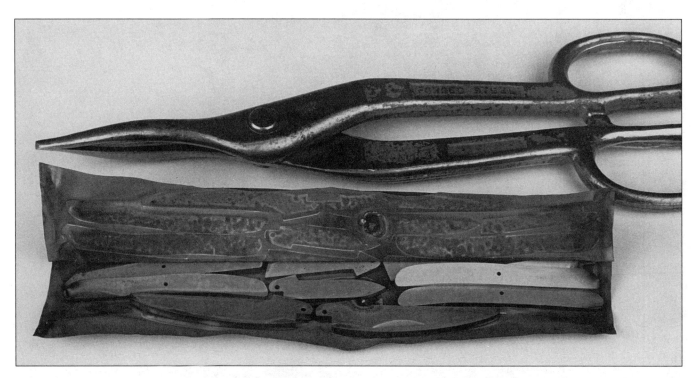

*After heat-treatment cut the foil packet open. Watch out for those sharp edges!*

blades I make are thin enough to be air quenched. Thicker sections of ATS-34 must be oil quenched to cool quickly enough to harden. Oil quenching steel is really messy and smoky!

As soon as the foil packets have cooled so they can be handled, I cut them open very carefully with tin snips. I swear the foil is even sharper now!

The blades and springs are checked against a flat surface and any minor straightening is done. After being heated to the high end of its hardening range, ATS-34 is easily straightened. A file won't make a mark on it, but the steel has not yet completely hardened all the way through. This is a characteristic of many stainless alloy steels. Some are "usable" without a subzero quench (such as 440C); others like ATS-34 are not, in my opinion.

When a test coupon of ATS-34 is broken before the subzero quench, the grain of the metal will look fairly coarse with jagged, torn areas. This means the steel has not completely converted to martensite, its hardest crystalline state, even though it outwardly appears to be hardened.

The subzero quench is essential to completely harden the blades and springs. Correspondence with Hitachi, the maker of ATS-34, indicates that -90°F will complete the hardening. It has to be at least that cold! My own early experi-

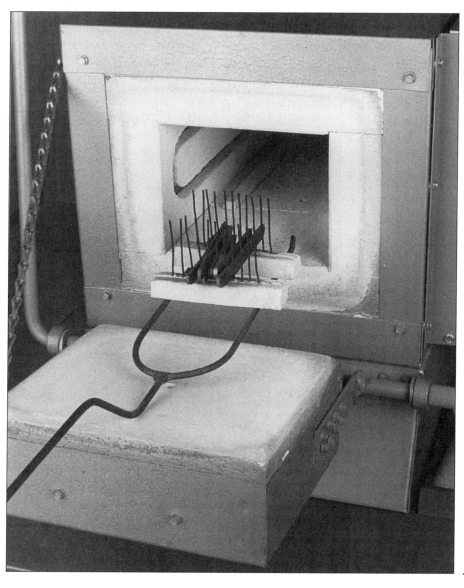

*After the subzero quench, place the springs back in the furnace for tempering.*

ence was that not even long periods in the family deepfreeze would get the job done.

My shopbuilt subzero freezer goes down to -180°F and does a fine job. The knife parts are hung on a rack in the freezer for three hours and then carefully taken out and allowed to warm slowly to room temperature. They are as brittle as glass when taken from the freezer. If handled with bare hands they will stick to the skin, resulting in a frostbitten spot. Compare this to dry ice (solid $CO_2$) which has a temperature of -109°F.

When the blades are back at room temperature they are ready to be placed in the furnace again for tempering. At this point the steel will be at a hardness of about Rc 62-63 (the Rockwell C scale is a standard measure of metal hardness). Test coupons at this stage are very brittle, displaying a very fine white grain structure when broken.

ATS-34 blades can be tempered at a low temperature of about 400°F or at a much higher temperature of 975°F. I believe that tempering at the higher temperature of 975°F results in a tougher, more flexible blade. As it is heated during tempering, ATS-34 at first softens and then starts to get harder again as it nears 975°F. At this temperature the hardness will be about Rc 60-61.

Coupons tested after tempering are noticeably harder to break and show a fine, slightly gray grain structure. The amount of toughness in the coupons after tempering is surprising.

I temper the blades for two hours at 975°F, then take them from the furnace and allow them to cool to room temperature. Then they go back in for another two hours at the same temperature. It is not necessary for them to be enclosed in heat-treating foil for tempering even though they will be at a very low red heat.

The long periods of time are needed for tempering because all of the stainless blade steels I have knowledge of change

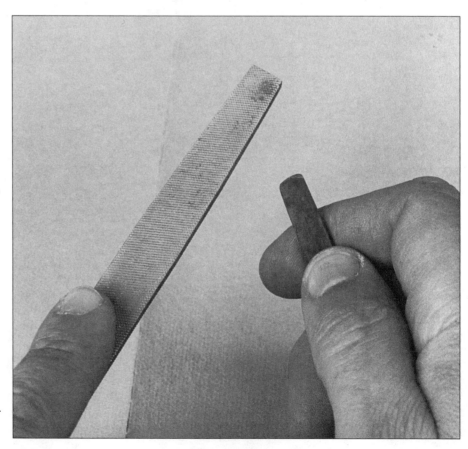

*Test the springs with a file for proper hardness.*

their grain structure quite slowly when heated. This is a big difference from the straight carbon steels which change very quickly as a general rule.

There seems to be a lot of mystery surrounding the making of springs. The fact is that it is not that difficult after a few basics are mastered. The hardness of the spring is its most important quality. A spring that is too hard is prone to break. A soft spring will not hold its shape after being flexed. Most of the skill in making springs is in learning how hard they should be. The remainder is in knowing how they should be shaped. Experience is the best teacher and that is only obtained by making springs and testing them.

I temper the springs in the furnace at 1025°F for two hours, then take them out and allow them to cool to room temperature. They are placed back in the furnace at 1050°F for another two hours. After being removed from the furnace and allowed to cool, I test them on each end with a worn 6 inch mill file. A new file will cut the steel when it is slightly harder and will give a faulty indication, but after a few trips over the hard steel, the new file quickly becomes a worn file. For my test the file should just barely cut the steel. However, at this point in the spring tempering, most of the springs will still be so hard that the file will slide on them. The ones that the file cuts are set aside and the remainder go back in the furnace at 1070°F for fifteen minutes. This procedure is repeated, with return trips into the oven at 1070°F for fifteen minutes, until the file has cut on each end of each spring. The Rc hardness on these springs will be about Rc 50. The procedure I use turns out a good spring every time for me. I feel it's a more reliable method than one strictly based on temperatures and Rc readings. Each spring gets exactly the right temper this way. I've found that ATS-34 from bar stock and from plate will temper at a slightly different temperature even though they are of the same composition.

Extreme care needs to be taken with the spring tempering time at the higher temperature, because the hardness of ATS-34 drops off very dramatically after it passes 975°F. The temperature-versus-hardness curve looks like it fell off a cliff; it goes almost straight down! With these conditions it is very easy to get the springs too soft. I like to leave the spring as hard as possible while retaining adequate flexibility. This eliminates as much wear as possible between the spring and the tang of the blade. Galling at the tang of the blade, which is always a concern when making slip-joint knives with stainless steels, is also minimized. Galling is the tendency of some metal surfaces to stick to each other when under pressure. The usual result is that the softer surface will have a portion torn away when the pieces are moved. When this happens to a spring, it creates a rough spot in the action of the blade.

I hope that this section will encourage you aspiring knifemakers out there to try heat-treating your own steel. Even if the equipment requirements prevent you from working with a steel like ATS-34, you'll find that a nonstainless carbon steel like O-1 requires a minimum of experience and equipment. Good old O-1 is inexpensive, easy to work with, and can make a knife at least the equal of the pocketknives your dad or grandpa carried.

I learned to heat-treat steel many years ago. Without that knowledge I doubt that I would be making knives today. It provided an essential starting point.

OK, I know there are those of you out there who just want to get on with making knives. For you, we've included the names of commercial heat-treaters in the listing of suppliers in the back of the book. They all do a great job.

## Bone Jigging and Dyeing

Jigging and dyeing the bone for the handles is easily done while heat treating is in progress. There are many reasons for doing the jigging and dyeing myself. Really good quality jigged bone is getting increasingly hard to find. Much of the current production lacks the texture and nice color of the antique factory bone. Jigging patterns on the commercially available bone are too regular and too large for my taste. Also, almost all the current bone is flat which makes for a slab sided handle. A little roundness in the jigged area can make all the difference in the world in the feel of the knife.

With that said, let's jig some bone (by the way, the "jigging" I talk about here has no relation to the "jigs" used earlier to form knife parts. Isn't English a confusing language?). The white bone that is the raw material can come from many sources. I have used camel bone from Turkey, mastodon ribs, and elk bones from my occasional hunting successes. Last, but not least, are the suppliers of handle material who increasingly stock plain white bone slabs. Whatever the source, the bone should be clean and white before starting. If the raw material is really raw it should be soaked in a strong detergent solution and then in bleach so that nothing but clean bone remains.

Starting with several sawn slabs of clean bone, draw a slightly oversized handle outline on the bone and grind the excess away with a sharp 60 grit belt. Here, as in most other operations in knifemaking, heat buildup has to be avoided. We'll do a number of handles at once because the individual pieces of bone will all take the dye a little differently. The more pieces there are, the better the chance for a good color match.

Stick a piece of tape, folded back on itself so it has a handle, on each piece of bone. Flatten one side on the disc grinder

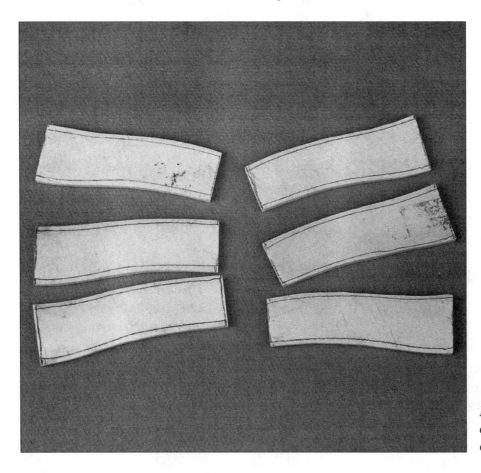

*Mark a slightly oversize handle outline on the bone slabs and grind to shape.*

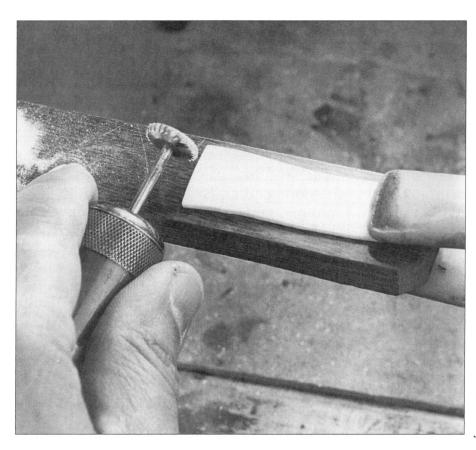

*Jig the bone by tapping it with the rotating cutter. Try to achieve a random pattern of jigging.*

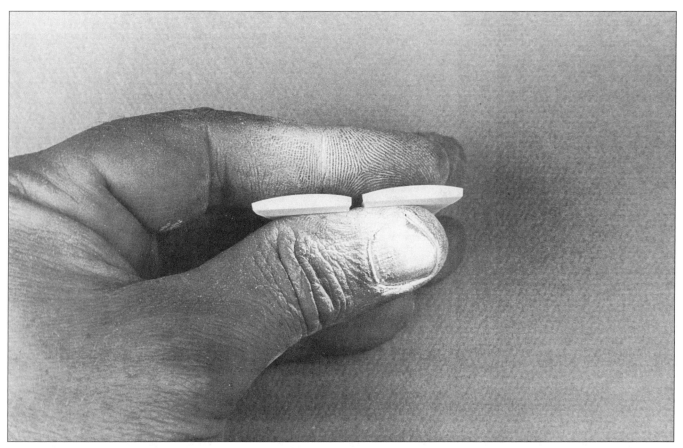

*Grind a rounded surface on what will be the outside of the jigged bone handles.*

129

using 120 grit paper. Remember that an equal number of "lefties" and "righties" are needed. Now, using a new 220 grit belt on the belt grinder, round the handles on what will be the "out" side. The final thickness of the bone should be just less than 1/8 inch at the center. Excess thickness will be removed from the back of the bone pieces.

I do the actual jigging with a small wheel-shaped cutter (MSC #00439158) in my flex shaft tool. A Dremel® tool works just as well. I usually clamp a small piece of board in my blade sanding vise so it is at a comfortable height and then clamp the bone piece to the board with a spring clamp.

To cut the jigging, bring the cutter up to maximum speed and just tap it on the bone. No special effort is made to space the jigging cuts exactly in the style of jigging I usually do. I like a random pattern that completely covers the rounded surface of the bone. A light positioned so that it reflects off the rounded surface of the bone allows me to see the areas I've missed. With this size knife I like to use a fine jigging pattern, so I use small light taps. On a larger knife, like a Sunfish, the taps are harder and of slightly longer duration to make a deeper and coarser pattern.

After jigging all the pieces, the dyeing of the bone can start. I use a combination of commercial leather dyes in a mason jar to soak the bone. Leather dye is extremely flammable due to the solvents involved and has to be used carefully. I *strongly* recommend doing any bone dyeing outdoors.

I use a mixture of a heavy dark brown harness dye with a lighter shade of regular leather dye to give a dark outer color and lighter inner shade. Tandy and Weaver Leather stock the dyes I use. It takes a lot of trials on test pieces to come up with the right combination of color and time in the dye. Keeping a written record

*Use leather dye to color the bone pieces. This oil dye does not contain oil.*

*Hang the dyed bone pieces in a warm place to dry thoroughly.*

of time and dye used helps get consistent results when the next batch is done.

Drilling a small hole in an excess area of the bone and stringing the pieces on a fine wire allows the pieces to be removed and dried easily. It will take several days in a warm place for the last of the dye solvent to evaporate from the bone. Then take a rag and a little acetone and wipe the bone to remove any excess dye on the jigged surface.

The jigged bone pieces are now flattened on the disc grinder with 120 grit paper and brought to the proper thickness for this size knife, which is about 3/32 inch, at the center of the bone pieces. After grinding, set the pieces aside for a day. Patience is a virtue in short supply in my shop, especially when I just want to get on with the knife, but I've found that giving the bone a chance to adjust after being sanded results in fewer problems with the handles later. Place the pieces gently back to back the next day and hold them up to the light. If they are still flat the handle fitting can begin, otherwise they are carefully flattened again on the disc grinder.

Fit the jigged bone pieces to the milled handle sides by carefully grinding the surface that contacts the front bolster. The work table on the disc grinder is set at exactly a 90-degree angle to the disc surface. Hold the jigged bone against the front bolster and over the top of the rear bolster, checking for alignment. The jigged bone should be roughly parallel with the handle shape and only have about 1/8 inch left to grind off to fit the rear bolster.

The jigged bone is now ground at the rear bolster so that it will just start to slip down between the bolsters from the top. Sighting down the line between the jigged bone and the bolster makes it easy to see which area of the bone needs to be ground off to let it slide further down. Use a magnifier to check the fit as the jigged bone reaches its proper location.

With the jigged bone fitted between the bolsters, hold the handle up to the light and try to see between liner and bone. If

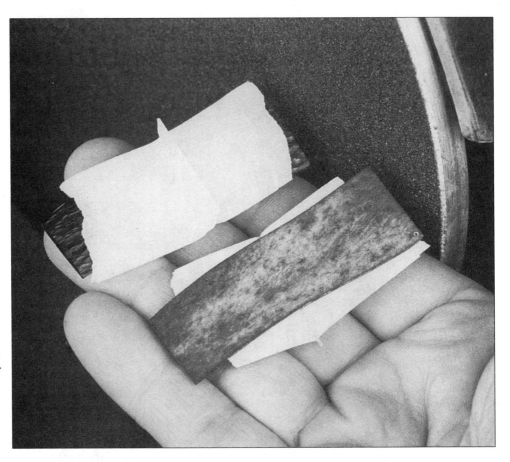

*Grind the jigged bone pieces to a thickness of 3/32" for this knife. Tape, folded with a "handle" in the middle, saves your fingertips.*

light can be seen, it means something is holding them apart. A little checking and some time with a small file will find the high spot. Fit up the other handle in the same way.

Now that the bone has been fitted in both handles, they are ready to be glued in place. I use a structural adhesive, Loctite Speedbonder®, that has much better adhesion than epoxy. This is a tip I received from knifemaker "Lee" Thompson, and it's a good one.

Choose a piece of 1/2-inch-thick Micarta™ that is smooth and flat and wrap waxed paper around it. Clamp the Micarta™ in the vise at a convenient working height, laying the jigged bone pieces and liners out on it. Spray the handles with Loctite activator, gently shake them off, and clamp on both bolsters with spring clamps near the edge of the Micarta™. They are placed so the jigged bone pieces tighten as they slide toward the edge of the Micarta™.

Apply and spread a thin layer of glue on the bone pieces with a razor blade; don't forget the ends. Because the glue doesn't set until it contacts the activator, there is no special hurry until the bone pieces are placed in position. The glue then sets in about twenty seconds. I place the jigged bone pieces in position, slide them down tight, and clamp them with a spring clamp on each end. If the phone rings, I ignore it! I tip the clamps up slightly just before the spring pressure is applied so the clamps will tend to pull the bone pieces tighter against the bolsters.

Allow the glue to cure overnight while the handles remain clamped on the Micarta™ sheet. After removing the clamps, peel the liners away from the wax paper and clean excess glue off with a little acetone on a paper towel.

Using a new 60 grit belt I grind away the excess jigged bone to the profile of the handle, being careful not to grind into the metal.

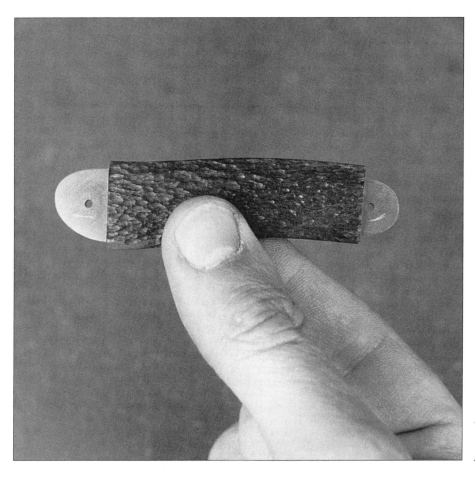

*Fit the jigged bone to the front bolster so that the bone will be parallel to the handle.*

*Grind the bone to fit the rear bolster and check with a magnifier.*

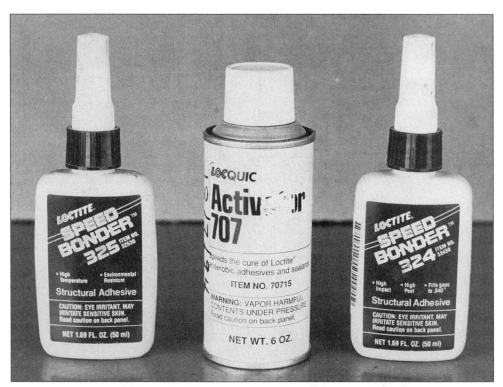

*This adhesive is stronger than the handle material, but you must be able to work fast.*

*Clamp the glued handles on a flat sheet of Micarta™. Waxed paper keeps them from sticking.*

*Pinning the hardened handle pattern to the handle and drilling in a holding fixture assures that all holes are accurate.*

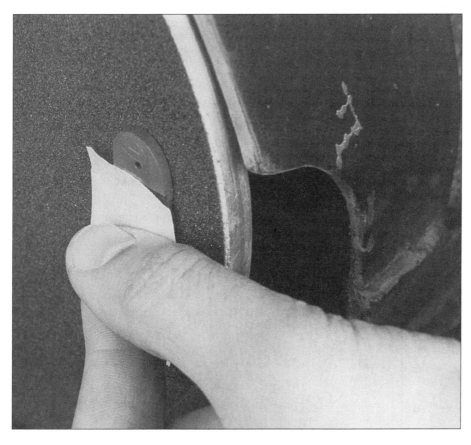

*Lightly grinding the back of the handles removes any glue residue and flattens the inside of the handles.*

*Keep the handle sides clamped back-to-back when not working on the knife to keep them flat.*

Then, after putting the hardened handle pattern on the liner side of the handle and placing #48 pins in the pivot holes, I place this assembly, liner side up, in a shop-made drilling jig and clamp it. These precautions assure that the holes are drilled square to the handle and in the right place. Using sharp drills of the proper size for each hole, I drill the handle material and center pin holes carefully through the bone on each handle.

Flatten the back of each handle on the disc grinder using 120 grit paper. This levels the liner and removes any glue residue. I always keep the two handle sides clamped together when I'm not working on them, which helps keep them flatter than they would otherwise stay. Temperature and humidity swings cause the bone to shrink and swell, warping the handles. Clamping them back-to-back resists this annoying tendency.

This completes the handle fitting.

## Setting up the Knife

It's finally time to start making something that looks like a knife. Let's take a little time and look at what we'll be doing during the knife set-up.

Grinding the blades and springs with my belt surface grinder gives me flat, clean parts to work with. Using the hardened handle pattern to accurately drill holes, a set-up block is made next. This block allows accurate fitting of the springs. The springs are scribed and then ground to bring the blades to the proper open position. Following this, the first trial assembly of the knife allows the back of the springs to be dressed off. Then the inside of the springs are ground to give the proper working tension when the knife is assembled the second time. Making the tapered center divider lets the knife be assembled once more, and allows it to finally start looking like a whittler. Whew! Nothing to it. I'll fill in the specifics.

*Surface-grind the small blades and springs to a thickness of 0.70".*

First, I surface-grind the heat-treated parts to proper thickness. When using my belt surface grinder I start with a 60 grit belt to remove the majority of the metal, followed by a 220 grit belt, and then a 400 grit belt to finish the parts. To eliminate the "bump" that would occur otherwise with tape splice belts, I grind the grit and binder off the belt at the splice.

The springs and small blades are surface ground to a thickness of .070 inch. I try to remember to take the majority of the excess thickness of the small blades off of the side opposite the nail nick. Place the parts closely on the magnetic chuck and surface grind them all at once. Turn the small blades as necessary, taking an equal amount off each side of the springs.

The two surface-ground springs are placed together and measured with a micrometer. To that measurement—.140 inch—an extra thickness of .004 inch is added because the two springs on a whittler come together at an angle on the back square of the main blade. This means that they are slightly apart over the pivot hole. The main blade must be about .004 inch thicker than the sum of the springs' thicknesses to allow for this. Taking an equal amount off each side, grind the main blade carefully to .144 inch.

*Surface-grind the main blade to a thickness of 0.144".*

Before going any further, grind the round ends of the blade tangs on the disc grinder with 220 grit paper. Using a small metal table with a #48 pin, bring each blade in contact with the disc and pivot the blade to grind an even radius. The finished radius is 0.10 inch for all blades. I remove the grit marks by buffing the round end of the tangs with a hard felt wheel using Brownell's 555 compound. It is important to keep all the tang edges square and sharp.

Use a piece of 1/2-inch-thick Micarta™ or Pakkawood to make a set-up block for marking the springs. The ideal size is about 8 inches x 8 inches. This allows several other knife styles to use the same sheet.

Start by placing the hardened handle pattern on the Pakkawood sheet, leaving room for the blades on each end. Clamp the handle pattern and drill the pivot and center pin holes to the proper sizes. Then, after inserting pins in the pivot holes,

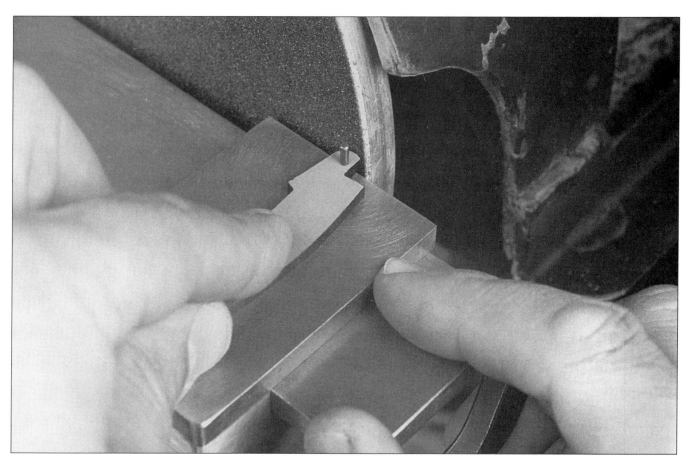

*Grind the round end of the tang to its finished radius.*

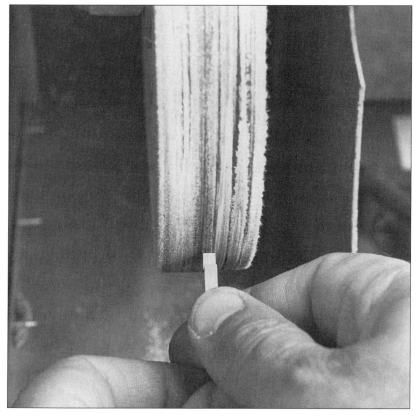

*Buff the round end of the tang, keeping all the edges square.*

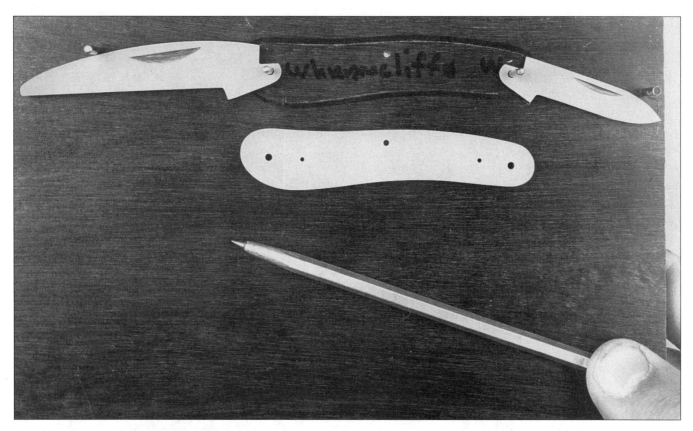

*Use a piece of Micarta™ or Pakkawood to make the set up block for spring fitting.*

scribe around the pattern. This identifies the knife by its outline. Remove the hardened pattern and place the blades over their pivot pins in the set-up block and align them as they should look when they are in open position. Clamp and scribe around each blade. Drill a 3/32-inch hole just on top of the scribed line near the tip of each blade for a stop pin. This completes the set-up block. We're ready to begin fitting up the knife.

Place the pivot pins and a center pin in the holes of the set-up block. Grind the ends of a spring enough so it will fit on the center pin with its ends on top of the pivot pins. The main blade is placed over the main blade pivot pin and held in place against the stop pin at its tip while scribing the outline of the back square onto the spring below.

Scribe both springs on both ends and then take them to the belt grinder. Grind the ends to within 1/16 inch of the scribed lines. Grind the areas of the springs that will rest on the blades to within about 3/32 inch of the scribed lines.

Clamp the springs together with a #52 pin in the center hole. Grind the "walk" area of the springs with 120 grit paper to square them with the sides of the springs. Grind one end in this way to within .035 inch of the scribed line. This establishes the "preload" that will give the springs the tension to hold the blade open and closed. Finish the walk surface with 400 grit paper on the disc grinder.

Placing the spring on the center pin in the jig and the main blade on the pivot pin, press up on the rear end of the spring. The blade should try to go into its open position, but not quite make it. Looking at the spring where it contacts the run up on the blade should show where the front of the spring needs to be ground to let the blade move up higher. When the

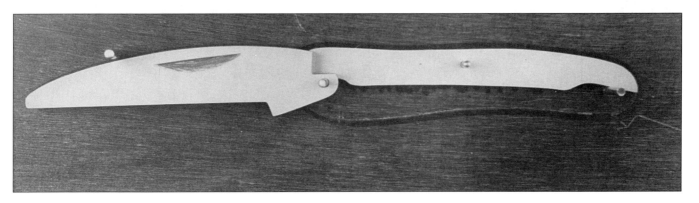

*Scribe the outline of the back square onto the spring below.*

Grind the walk area of the spring to within 0.035" of the scribed line. This provides "preload" pressure.

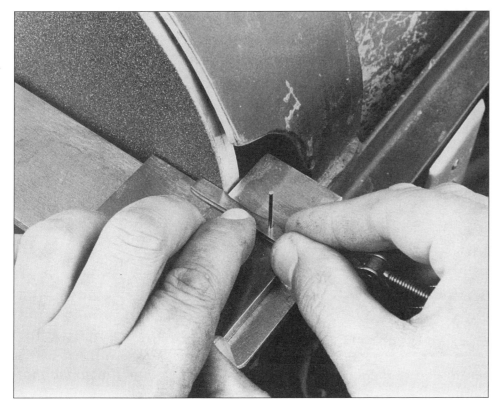

*Clamp the springs together and grind the walk area on the springs to establish the "preload," which is necessary to hold the blades open and closed.*

*Grind the end of the spring to allow the blade to go into the open position. Remember, metal is easy to grind off, but impossible to put back on.*

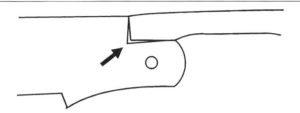

Grind the spring so it contacts the tang at the top of the run up and the rear edge of the back square. This leaves room for pocket lint to collect.

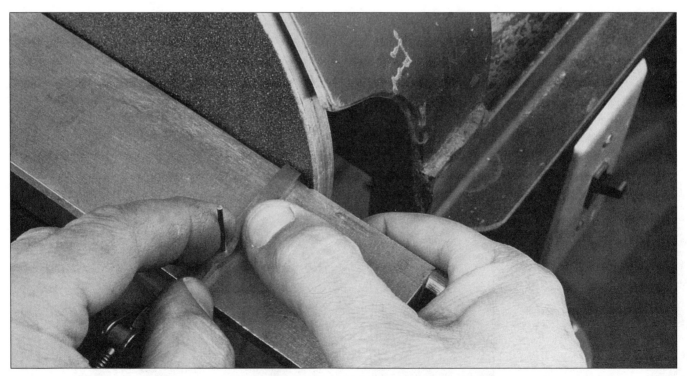

*Clamp the two springs together and grind the longer unfitted one to the length of the spring that was fitted first.*

blade is in its full open position the front surface of the spring should be in contact at the top of the run up, but not at the bottom. This assures a nice tight line between the spring and the top of the blade when the knife is finished, but it still allows space for the inevitable pocket lint to collect.

Once the blade is very close to being in full open position, clamp the spring that has just been fitted to the top of the other spring, placing a #52 pin through the center holes. Grind the longer spring on the bottom until the end of the top spring almost touches the disc. Then, with the springs still clamped, finish the ends of both springs on the disc grinder with 400 grit paper. Check the springs again on the fitting block to make sure the blade is in exactly the right position. If the springs are ground too short by mistake, the careless knifemaker gets to heat-treat replacement springs!

Once the springs have been ground to proper length at the main blade end, use the same procedure to fit up the small blades.

The first assembly of the knife will allow us to shape the springs on the back of the knife. Since the springs have been identical to this point, there has been no need to mark them. Now that they will be shaped with the knife, that will change. Using a vibrating engraver, I mark a P on one spring for PEN blade and a C on the other for CUTOFF pen blade.

Placing the center pin in the pile side handle, I stack the C spring and then the P spring on it, followed by the mark side handle. With the main blade in position, insert the front pivot pin. With the small blades each in their proper position and pushed up tight against their springs, work a #60 pin all the way through the rear pivot holes. The springs are still far too stiff in their unshaped condition to allow the assembly of the knife with both #48 pins.

With a sharp 60 grit belt on the belt grinder, I use the contact wheel to grind the springs down to the top of the handle profile. This is done ONLY in the area from the center pin to the main blade. Now reverse the assembly process and place the #48 pin through the small blades at the rear pivot hole and use the #60 pin to hold the main blade in place. Grind the springs down to the handle profile from the center pin, back to the small blades.

Now that we have profiled the top of the springs, we don't have to worry about

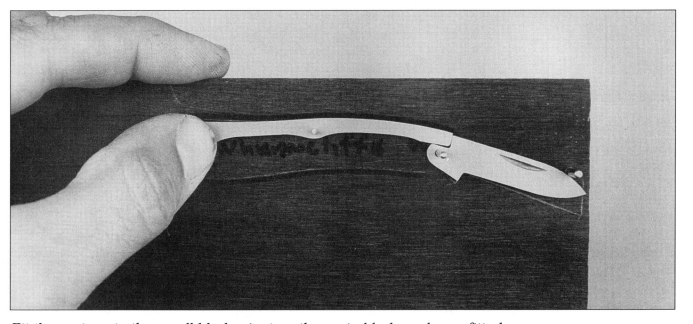

*Fit the springs to the small blades just as the main blade end was fitted.*

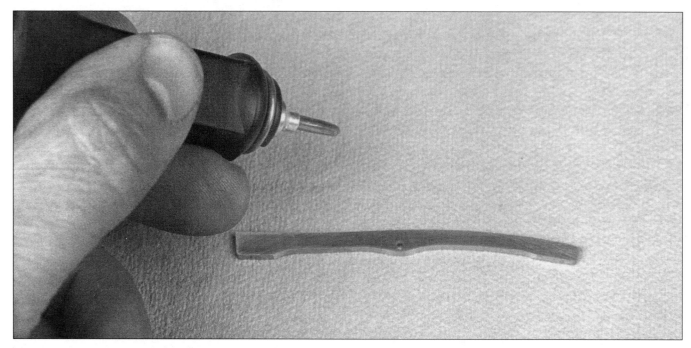

*Mark the springs for identification before shaping the top of the knife.*

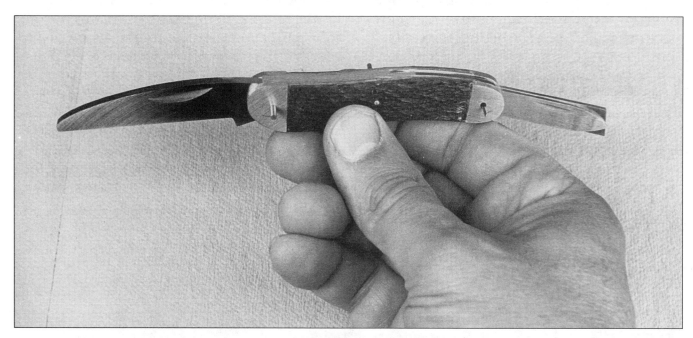

*Assemble the knife with a small pin in the rear pivot hole and grind the front of the springs to handle profile.*

grinding too far on the underside and making the springs too weak. Using the 1 inch contact wheel on my small radius grinder and a 220 grit belt, I grind the underside of the springs to shape. I always leave the springs a little strong until I've finished the back of the knife and worked the action a dozen times or so. The photo shows both blades assembled with a spring on the setup block. This illustrates how the spring should be shaped.

Before continuing work on the springs, we'll make the tapered center divider. Because the divider is tapered, the jig that holds it for grinding is also tapered. Cut a piece of 3/16-inch steel about 1 inch wide

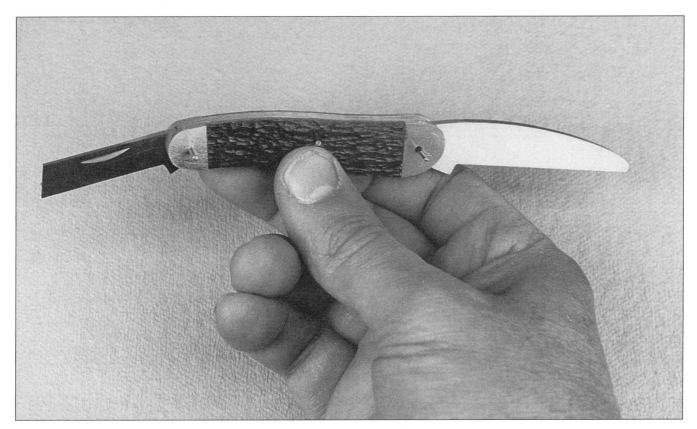

*Reassemble the knife with a small pin in the front pivot hole and grind the remainder of the springs to handle profile.*

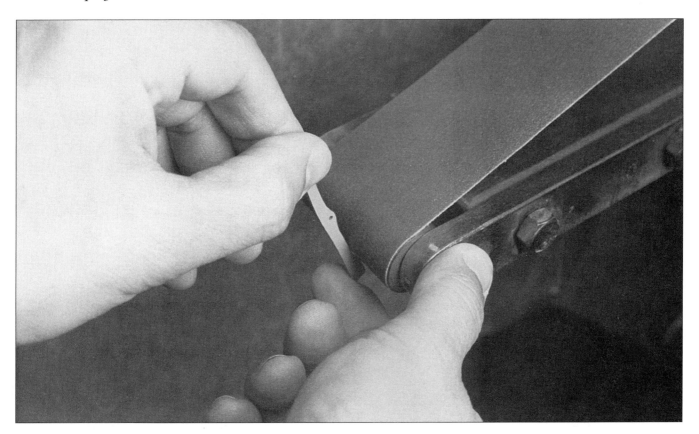

*Grind the underside of the springs to shape.*

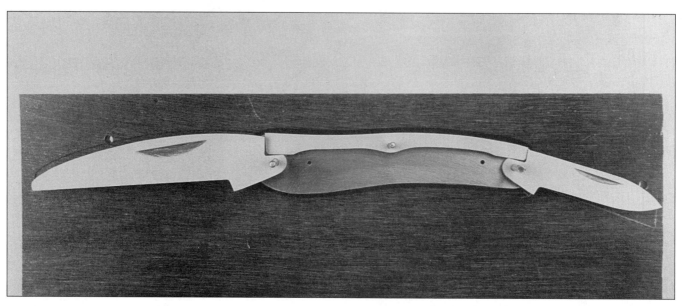

*The blades and one of the springs have been assembled on the set up block to show how they fit together.*

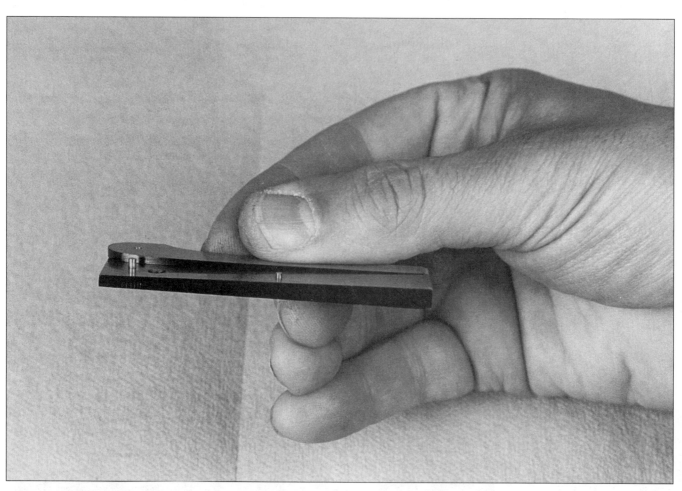

*The tapered divider (on top) and the divider grinding jig show their respective tapers here with the help of a little black ink on their edges.*

to 3-1/4 inches long and surface grind it on both sides. Place the hardened handle pattern on it. Then clamp and drill the center pin and rear pivot holes. Placing one of the springs over a pin in the center pin hole, mark the front end of the spring ends and cut off the jig at this point. Now we go to the surface grinder. Because the divider will be .080 inch thick at the back of the handle, the jig is blocked up that amount at the rear. The magnetic force is applied and the jig is ground flat. A #52 pin goes in the center pin hole and a #48 pin in the pivot hole of the side of the jig opposite the side just ground. Hold the pins in place in the holes with a drop of Super Glue™.

Cut the divider blank from a hardened piece of 416 stainless 1/8 inch x 3/4 inch x 3 inches. Surface grind both sides, scribe the handle outline, and then grind to shape. Drill the center pin and rear pivot holes.

Back at the surface grinder, place the divider blank over the pins in the jig, apply the magnetic force, and grind the divider. When the divider has been ground to a knife edge at the thick end of the jig, it is finished. I carefully straighten the divider if it needs it. The thin edge of the divider is ground back to an area where it is .010 inch thick. If the paper thin edge is left, it will rise above the springs and cut a finger sooner or later. I know, because my first whittler cut the daylights out of me!

With the divider ground to a taper, it is next ground to shape on the area inside the handle. I use the O-1 blade pattern to mark the divider, allowing a little clearance with the tip of the blade.

Before the knife can be assembled with the divider in place, the pivot holes must be reamed to a slight angle. I accomplish this by first making a tapered block from a piece of Pakkawood 3 inches long. Because the spacer is .080 inch at its thickest point, the block is cut to a taper of one half that amount. From the center line of the knife, each handle is only .040 inch wider at the back than at the front. Grind the Pakkawood block on the disc grinder so that it is .040 inch thinner on one end than the other.

Now, with a #48 reamer in the drill press, place a handle, liner side down, on the tapered block. Place the main bolster at the thin end of the block. With the drill press at a medium speed, bring the turning reamer down so it lines up exactly with the main pivot hole, and ream the hole.

Ordinarily, enlarging a hole at an angle like this would not be an acceptable practice. In this case, however, the angle is only three quarters of a degree. This is not much of an angle. I've never had any

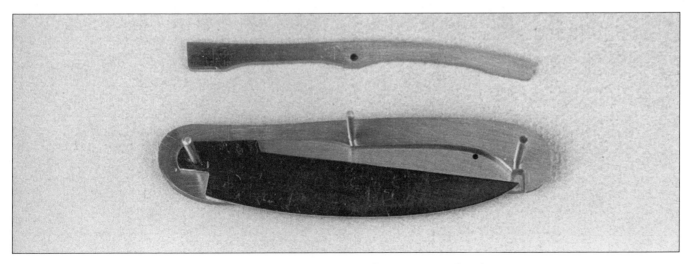

*Use the O-1 blade pattern to help profile the inside of the center divider.*

*Use a #48 reamer and the tapered block to give the pivot holes the angle required in the finished whittler.*

*Assemble the knife for the first time with #48 pins in both pivot holes. This gives the springs the same tension they will have in the finished knife.*

problem while putting a knife together after doing this. The center pin holes can be worked to a slight angle with a #52 drill.

With all of the holes "angularized" (I made that one up!), the knife can finally be put together with full tension on the springs, just as they will be in the finished knife. Stack the knife parts in proper order on the center pin. Put the small blades in place and insert a pin in the rear pivot hole. I now take the knife to the assembly press and, placing a bar on the top of the handle at the forward end of the springs, I push the springs up enough to insert a pin in the main blade.

Now that the knife is together, I put a tiny drop of Break Free Teflon Oil at the tang of each blade. This little precaution will help prevent spring galling the first time the blade is moved from open position.

I grind the back of the knife to profile, first with a 60 grit and then a 220 grit belt on the belt grinder. The back of the knife should have a clean surface all the way across the handles, springs, and divider.

Once I'm satisfied with the top profile of the handle, I try the action of the various blades. At this point they should all be just a little stiff. Using a small pair of Vise Grip pliers, I pull the main pivot pin to

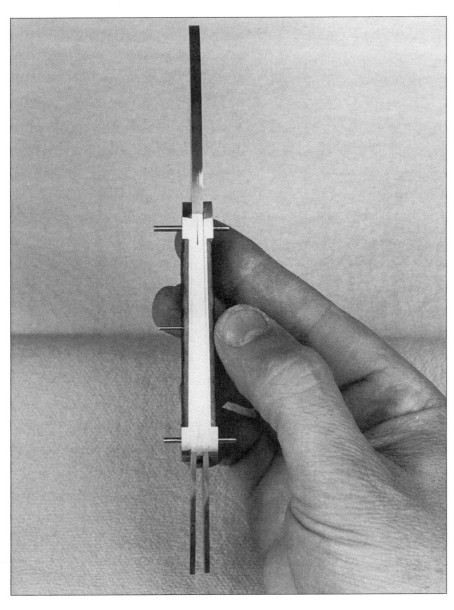

*The top of the knife is shown with the tapered divider in place.*

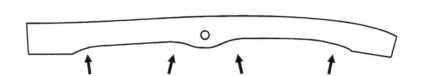

All curves on the bottom of the spring should be smoothly rounded.

release the spring tension and take the knife apart so the springs can be removed. If the springs need to be weakened a little, they are ground on the inside surface with a 220 grit belt on the small radius grinder. Care must be taken not to weaken them too much.

A few words on spring design are appropriate here. It is important that the bottom of a spring have a smoothly rounded profile. Because it is under tension, stress concentration has to be guarded against. If there is a scratch across the bottom of the spring, the spring will tend to break at that spot sooner or later. The same fate awaits a spring that is thinner in profile at the center pin than closer to the blade. The

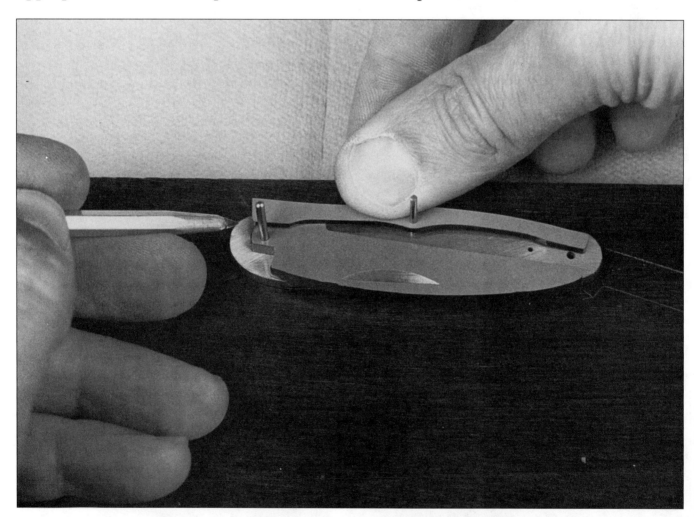

*Scribe the outline of the spring on the blade tang in the closed position. This will allow grinding the bottom of the tang to shape.*

bending force will tend to concentrate in that thinner area.

Assemble the knife again, now that the spring tension has been adjusted. Let's assume it's right on the money this time. The next step is to work the blades from full open to a half closed position a dozen times or so. This establishes a wear pattern on the walk area of the spring. It is important not to disturb this wear pattern, since once this has been formed, any tendency for the spring to gall is greatly reduced.

Now that the springs are adjusted, disassemble the knife again and finish the bottom of the springs with an 800 grit belt on the small radius grinder. The spring bottoms will be hand sanded through 1500 grit before the knife is assembled permanently.

I next bring the kick and bottom of the tang on each blade close to their finished dimensions. Leave a little extra metal for final fitting.

Place the pile side handle on the set-up block and insert the main pivot pin and center pin. Place the main blade over the pivot pin and swing it to an almost closed position in the handle. Place one of the springs over the center pin and hold it down over the tang of the blade. Raise the front of the spring slightly above the handle profile and scribe the outline of the bottom of the spring on the blade tang.

Take the scribed blade to the small radius grinder and grind the tang bottom

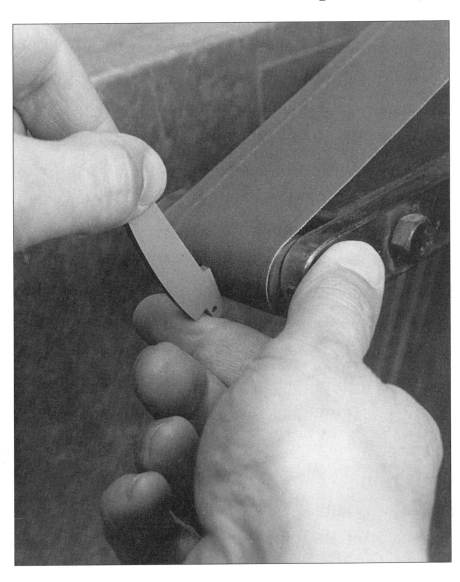

*Grind the bottom of the tang to its almost-finished shape. Leave a little metal for final fitting.*

Knifemaking with Terry Davis

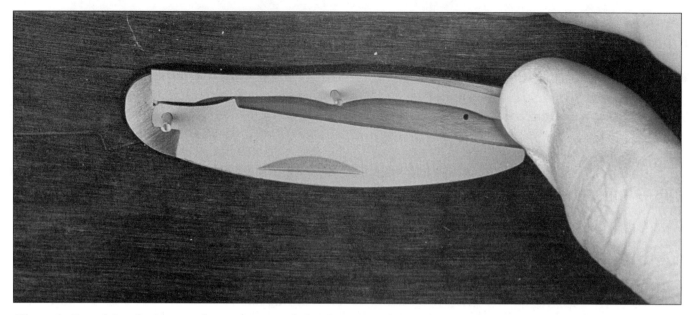

*The relationships between the spring and the bottom of the blade tang can be seen here.*

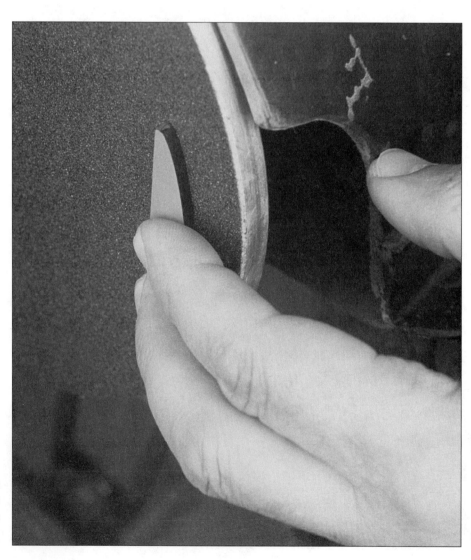

*Sand the sides of the blades lightly on the disc with 400 grit paper. Turn the discs by hand for control.*

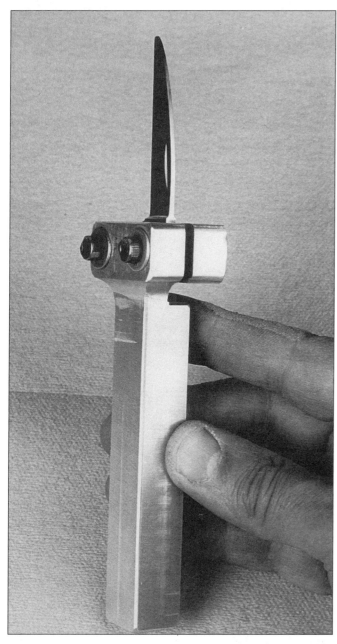

*This holder was machined from aluminum bar to hold the blades for grinding the bevels.*

to shape with a 220 grit belt on the 1 inch contact wheel.

Bring the tangs of the small blades to their almost finished shape with the same procedure used for the main blade.

I like to make sure that all surface grinding marks are removed from the sides of the tang area before blade grinding starts. Grind the sides of all of the blades with 400 grit paper on the disc grinder. Once the blade bevels have been ground, any heavy sanding on the small area of the tang could round off surfaces of the tang that shouldn't be rounded. In this instance, as in many others in knifemaking, the sequence in finishing parts is very important.

I clamp the blades in a holder that I machined from aluminum bar. This holder grips each blade securely for grinding and has a nice light handle for good control. It works much better than the Vise Grip pliers I used when I started.

I'm fairly conservative in my choice of belts for grinding blades. For rough grinding I use the Klingspor CS 310, a good quality aluminum oxide 60 grit belt. The longer lasting zirconia and "hogger" type belts just don't work as well for me. I like to grind with sharp aluminum oxide abrasive on a belt, even if the belt doesn't last as long as some of the others.

Grind the main blade first. I grind a step-down forward of the tang on each side to take the blade to a thickness of 0.10 inch. Next, the bevels are ground to an edge thickness of .025 inch. On a straight edge blade it is important not to grind the edge too thin at the tip of the blade. I remove the blade from the holder several times and place it between two flat pieces of steel to check that the point is centered. If it is not perfect, there is enough steel remaining to center it.

With the blade roughed out, I start finish grinding with a 220 grit belt. I use a very small amount of wax lubricant, Chem-Trend 140 Stick Wax, on the back of the 220 and finer belts to prevent little globs of a varnish-like substance from building up on the platen. These deposits cause a small depression to be ground in the blade bevels, ruining the blade. They will absolutely spoil your whole day. I have been known to throw blades out the shop door and into the driveway after these little globs have done their dirty work. I've used the wax lubricant for about nine years now without a problem, so it is safe to walk into my shop.

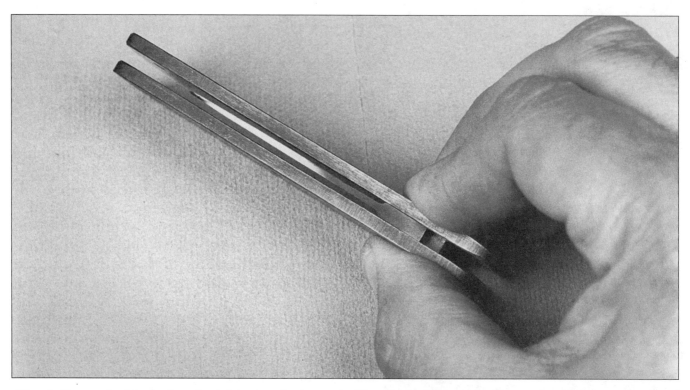

*Check the main blade for straightness during grinding by holding it between two pieces of ground barstock. Here the tip must be ground slightly on the right side to center it.*

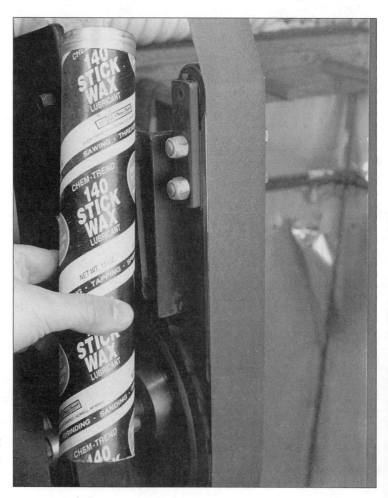

*A tiny amount of wax lubricant on the back of 220 and 400 grit belts eases the belt's trip over the platen.*

Using the 220 grit belt, I remove about .010 inch from each side of the blade to clean up the coarse grit scratches. Bring the edge of the blade to a thickness of .005 inch. This will make the edge easy to sharpen, and will help it cut much easier than a blade with a thick edge.

I use a 400 grit belt only to remove the 220 grit scratches and refine any areas of the blade that were missed by the earlier, coarser belt. A 400 grit belt is too fine to remove much metal. Years ago I used even finer belts while blade grinding, but I found that it was wasted effort since hand-sanding does a better job.

Grind the small blades next. With a 60 grit belt on the grinder, I start by grinding a .010 inch step-down on the side of the blade opposite the nail nick. This thins the blade for needed clearance with the main blade. Grind the bevels very carefully because there is not much metal to be removed with the coarse belt. As with the main blade, the edge should be .025-inch thick after the coarse belt grinding. When grinding the small blades for a whittler, they are offset-ground towards the nail nick side. The idea is to keep them as close to the knife handle, when closed, as possible.

*Clamp the blade in the sanding vise to hand-sand the bevels. Take the blade out of the vise every time you stop sanding.*

After checking their clearance from the handle in closed position, use the 200 and 400 grit belts to finish the small blades. Because these are very small, thin blades, great care must be taken not to overheat them or grind away the blade tips.

Then, after finish grinding all the blades, hand-sand them with Klingspor PS 11 400 grit paper. I use a piece of Micarta™ that has been flattened on the disc grinder as a backup block. All the edges on the block are ground square and have nice sharp corners. I cut the abrasive paper with an old pair of scissors to a width of 1-1/2 inches before use.

My sanding vise mounts to a bracket that holds it in a horizontal position just above waist level. This makes it very convenient for finishing blades, but VERY dangerous. If the phone rings or a neighbor drops by, I ALWAYS take the blade out of the vise. The image in my mind of what would happen if I ran into the blade while it was clamped in the vise makes me very paranoid. The closest hospital emergency room is twenty-five miles away!

Place the main blade tang in the vise clamped in a fold of heavy leather. Sand lengthwise along the blade, being sure to push firmly into the shoulder between the bevel and the tang. Care has to be taken not to cut or poke your fingers, because the blade, although not sharpened, will still cut. After a good bit of hand blade sanding you get a sense of when the block is flat on the bevels. If the block isn't flat, the edges of the blade will be rounded off, destroying the appearance. When the paper starts to dull, after maybe a dozen strokes, I slide it under the block about 3/4 inch to bring an unused area into contact. Always use sharp abrasive when sanding blades. Dull abrasive wastes your time and can give a poor finish.

When all the marks from the 400 grit belt have been sanded out on the bevel I'm working on, I change the direction of the sanding about 10 degrees. By doing this I will be removing the previous scratch pattern. This helps remove any lengthwise waviness in the blade. As soon as the scratch pattern is gone from the first sanding, go through the same procedure from the start on the other side of the blade. We're done for the time being with this blade.

The small blades are sanded using the same method as the main blade. Care must be taken with these blades, because they are very thin and could be broken if too much force is used on the sanding block.

The reason for sanding the blades now is that I like to remove any vertical scratches from the surface as soon as possible. Scratches in the blade act just like scratches in glass; they make it easy to break. "Stress concentration" are the key words here. This is the real reason knife blades are finely finished. The fact that they look pretty is just a bonus.

## Final Fitting and Blade Finishing

With the blades ground and rough sanded, we're ready to start doing the final fitting of the knife. This is the home stretch!

At this point I finish the bottom of the springs with 1500 grit 3M® 414Q Color Sanding Paper. I use a hard rubber eraser with one side rounded as a backup block for the sandpaper.

Assemble the knife and carefully dress the top of the springs with the blades open. Use a 400 grit belt. This will make it easy to see when the springs are not level with the top of the handle. I now close one blade at a time, making a mental note of what I must do to each blade. Every blade must close into the handle to a given depth and the spring must be level with the top of the handle, as it was when the blade was open. That's simple! Right? Actually, it's not so simple. When the kick

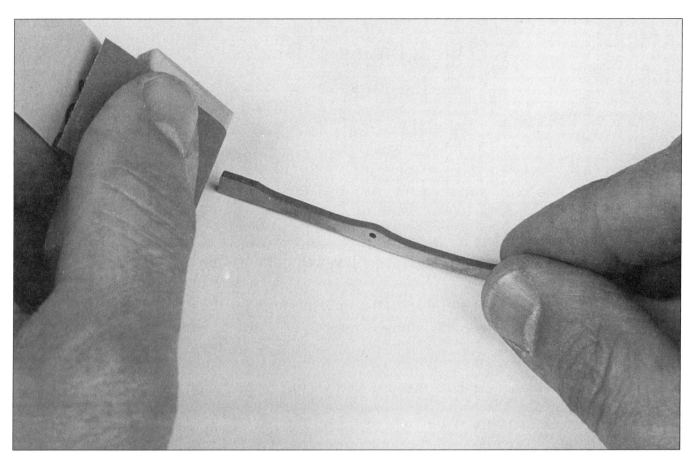

*Finish sand the bottom of the springs with 1500 grit paper.*

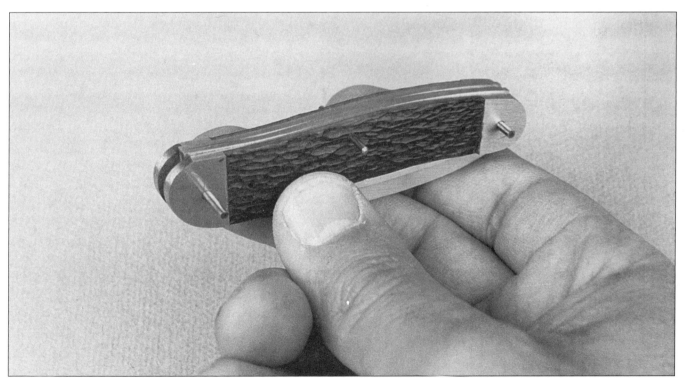

*The springs must be fitted so they will be level with the top of the handles when the blades are closed. Grind the bottom of the tang to bring them down.*

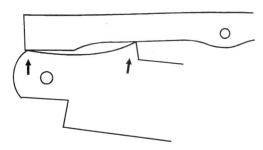

## Final Fitting of Blades

Remove metal from the rear of the bottom of the tang to bring the spring level with the handle.

Remove metal from the kick to lower the closed blade into the handle.

is shortened to drop the blade further into the handle, the spring will also drop down a little, and when metal is taken off the back of the tang to let the spring come down to the top of the handle, the blade will rise a small amount out of the handle.

To combat these tendencies I usually try to grind the kick to bring the blade to its proper position in the handle. Then I remove only a small amount at a time from the bottom of the tang, using the 1-inch contact wheel and a 400 grit belt. The kick is shortened slightly when needed to keep the blade in position as the bottom of the tang is ground. Most knives will have to be assembled ten to fifteen times in order to

*As seen here, the bottom of the blade tangs have been ground to bring the springs level when the blades are closed.*

*Use the disc to finish the front of the kick.*

get the springs level when the blades are closed. Remember, if too much is ground from any area, it's necessary to make a new part. This always happens to me when I don't have a heat treated spare part!

Counting the last assembly, our Wharncliffe whittler had to be assembled fourteen times to fit the blades in perfectly. That's about average! Whittlers have their own particular fitting complication: The main blade has to be perfectly level across the back square and also across the tang bottom. If either area is crooked, the two springs will not be at the same level when the main blade is closed.

Now, disassemble the knife for the last time. It's time to final-finish all the blades. We'll begin with the front of the kick on all blades. This is easily polished on the disc grinder with 1200 grit paper. Carefully polish the bottom surface of the tang with 1200 grit paper and the eraser back up block. All edges must remain sharp and clean.

Each blade is now clamped in the vise for sanding, this time by the blade end. Hand-sand the sides of the tang with 600 grit paper followed by 800 and 1200 grit. Be extremely careful not to round the tang surfaces. Now place the tang end of the blade in the vise. Sand the blade bevels next, taking extra care to push the paper into the rounded corner of the shoulder between the bevel and the tang. I usually make use of a magnifier to check this area for a clean sanding job. Sand the blade with progressively finer paper up to 1200 grit.

The top of the blades are finished on the disc grinder with 1200 grit paper. I like to slightly round the top of the blades and their edges. This gives a nice feel to the knife and, as long as I don't get carried away, the lines of the blade stay clean.

## *Trademarking*

To trademark my knives, I use the Kodak Thin Film Resist system. I get a nice mark this way, deep and very clean. The only downside to the process is that it is slow; however, because I am finishing other parts of the knife at the time, I'm never far from the workbench anyway.

This system uses a clear film positive to form a stencil on the blade. Acid is then used to cut the mark in the steel. This is the same process used to make computer chips, so duplicating fine detail is not a problem.

I make the positives using ortho-litho copy film in my 35mm camera. I use rub-on letters to make the original of the mark I want. I then take a picture of the original with my camera, develop the film to make a negative, and then use that negative to make the clear film positive. I've been told that the positives can now be done by computer, using a clear sheet with a laser copier. I hope to try that. The really nice thing about making the positives myself is that I can make them any size. The perfect = sized logo can easily be made to fit any knife I make.

The process used to mark a logo using the Thin Film Resist system starts with cleaning the tang with acetone. Apply a drop of resist. It should be about the same viscosity as homogenized milk. After rocking the tang back and forth to spread the resist, hang the blade up and let the excess resist drain off. Let it dry for about ten minutes, then place in a toaster oven at 170°F for twenty minutes to bake the

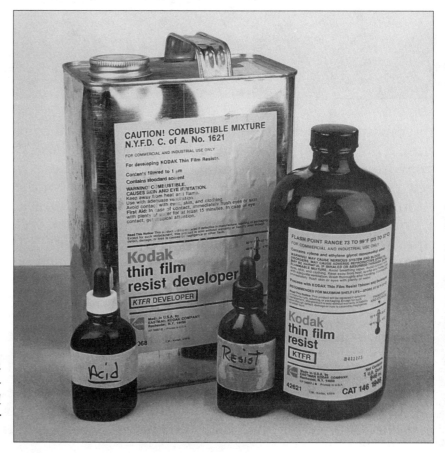

*Here are the components used in the Kodak Thin Film Resist marking system. The resist is thinned for use and placed in the dropper bottle.*

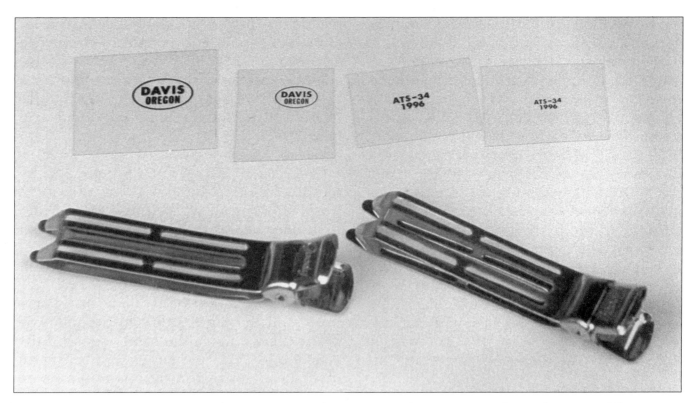

*Here are the positives used to form the logo. These are made to any size desired using a 35mm camera.*

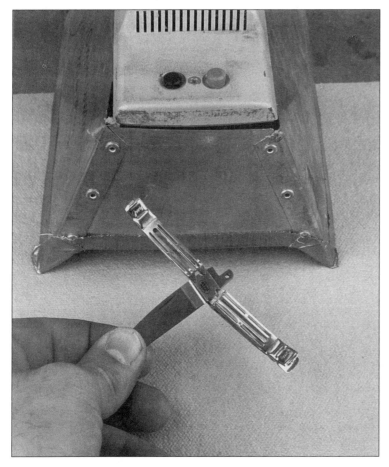

*Use hair clips to clamp the positive to the resist-coated blade tang and expose the resist to UV light for seven minutes.*

resist. Position the positive exactly where the mark is desired and use clips to hold it in place. Place the blade with the positive under an ultraviolet light for seven minutes. The resist under the clear portion of the positive will harden; the areas under the black parts of the logo will remain soft. When the time is up, I remove the positive and swish the blade around in the resist developer, which dissolves the soft areas of the resist. Now wash the blade under running water to remove the remaining developer. The blade, with its newly developed image, goes back in the toaster oven for ten more minutes to harden the resist. When the time is up, take the blade out, mask off the area of the logo, and use acid to cut the mark. I use acid rather than an electro-chemical etcher because it cuts the mark very quickly and does a beautiful job.

I apply the acid to the mark area from a two ounce bottle using an eyedropper. The acid used is a half-and-half mixture, by volume, of nitric and hydrochloric acids and is very aggressive when applied to the metal. (When mixing these two acids, it is necessary to go very slowly because a lot of heat is generated.) I strongly encourage anyone working with any acid to use safety glasses and rubber gloves and have a container of baking soda nearby to neutralize spills. I have a lot of respect for acid because of the damage it could do if I'm careless.

Once the acid has cut the mark to the required depth, which is about .005 inch, wash off the acid with running water and then use lacquer thinner on a rag to clean off the resist. The whole marking procedure actually goes a lot faster than it sounds.

I trademark the main blade on all my knives on the mark side. On the pile side of the blade I mark the steel used and the year made. I started serial numbering all my knives consecutively about ten years ago on the advice of knife collector Joe Drouin. This has worked well and makes my record book, with the design, handle material, steel used, and date, easy to keep.

All blades are now finished with 1500 grit paper and a hard rubber block. This

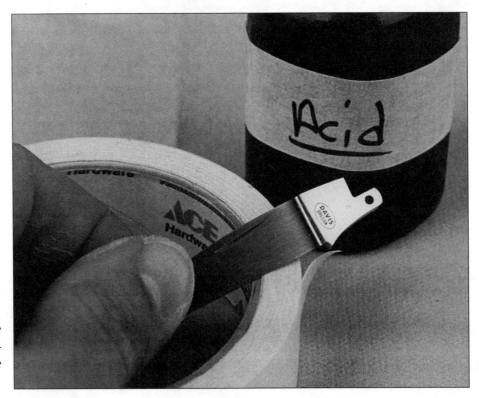

*Use acid to cut the mark. The resist acts as a stencil allowing the acid to eat away the exposed metal.*

block is made by gluing hard rubber to a piece of 1/4-inch Micarta™, which is trued and squared on the disc grinder with 120 grit paper. The hard rubber allows the abrasive to give a uniform scratch pattern, creating a nice sheen on the blades.

## Finishing the Handles

With the blades taken care of for the time being, we can start finishing the handles. The bolsters are contoured using a rotary jig to mill them to shape. The main advantage of milling, over simply grinding the bolsters, is that the mill generates very little heat while removing the excess metal. In years past heat would occasionally "knock off" soldered bolsters on small knives while I was intent on grinding. The heat can also cause pearl and jigged bone to develop milky-looking spots that ruin the handle, just when you're ready to put the knife together.

I built my bolster milling jig from pieces of precision ground steel. It is a near copy of one in Ron Lake's shop. I use a 1/2-inch extra-long milling cutter just for contouring bolsters.

Mill the bolsters to shape using an angle of 4 degrees to taper them toward the ends of the knife. Place the hole in the bolster to be cut on the jig pin and clamp the handle securely. Continue cutting the bolster

*Use the rotary jig to cut the bolster sides to rough shape. This method produces very little heat.*

*File the shield to shape using the hardened pattern as a guide.*

until the cutter just touches the jigged bone handle material. All four bolsters are done in the same way.

Both handles are taken to the belt grinder and finish-contoured with a new 400 grit belt. As always, be on the look out for too much heat buildup in the metal.

Making and fitting the shield is the next order of business. A knife like this wouldn't look right without a shield, but getting it installed cleanly can be a challenge.

For each knife size and type, I make several shield patterns and a corresponding jig to help cut the recess that the shield is installed in. This recess has to be nice and flat because I pin all my shields in, even the tiny ones. If everything isn't just right, about the second time the pin is struck with the hammer, the knife handle will crack. The resulting uproar usually makes our dog run and hide.

Make the shield first. I have used 410 stainless, sterling silver and occasionally gold for shields, depending on the customer's wishes. Because this is a special knife of sorts, let's use gold for the shield and pins. The material is 14 carat sheet in .050-inch thickness.

The first step is to drill the pin holes in the shield blank. For this I use a hardened strip of steel, with holes drilled at the spacing required by the various shields. Clamp the piece of gold sheet under the appropriate holes and drill the two 1/16-inch holes. Cut out the rough shield with a jeweler's saw and it's ready to file to shape. Insert two 1/16-inch pins through the holes in the shield pattern and blank and clamp in the vise. File to shape, being careful to keep the edges nice and square. Buff the edges of the shield on the hard felt wheel with 555 compound. Clean it up with a little acetone and it's ready to install.

I make a Pakkawood jig to hold the knife handle and the shield inlay pattern in alignment. With this setup I can put any shield I want in the handle and it will line up perfectly.

With the handle in the jig, ready to inlay the shield, measure down through the inlay pattern to the top of the jigged bone. To this measurement add .025 inch and set the depth of cut for the shield recess. The cutter I use for this is a small drill bit that has been reground to the shape of a little milling cutter. This cutter is clamped in the chuck of my flex shaft tool. Using this simple arrangement like a miniature router, I can cut a flat, clean shield recess.

I final fit the shield into the recess using my Optivisor and a few small chisels. The inlay pattern is purposefully made a little small to allow hand fitting of the shield. I scribe a small T, for top, in the shield so I always fit it back the same way. They are never truly symmetrical.

With the shield fitted up, clamp it in the recess with a spring clamp and take the handle to the drill press. Carefully drill both holes through the shield holes and on through the handle with a sharp 1/16-inch drill bit. Then, take the shield out and clean any shavings out of the shield recess and replace the shield. The holes in the shield are given a very light 20 degree countersink and checked with a magnifier to see that they don't have any chatter marks. The gold rod is cleaned up with 1500 grit sandpaper and inserted through the front shield hole and on through the liner. Now, cut off the rod, leaving a little

*Rout out the rough shield recess with the inlay jig.*

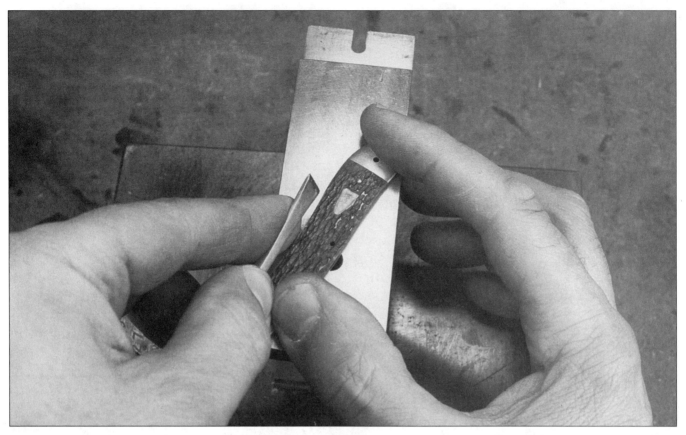
*Use a small, sharp chisel to finish inlaying the shield into the handle.*

*Peen the shield pins very carefully. It is very easy to crack the bone.*

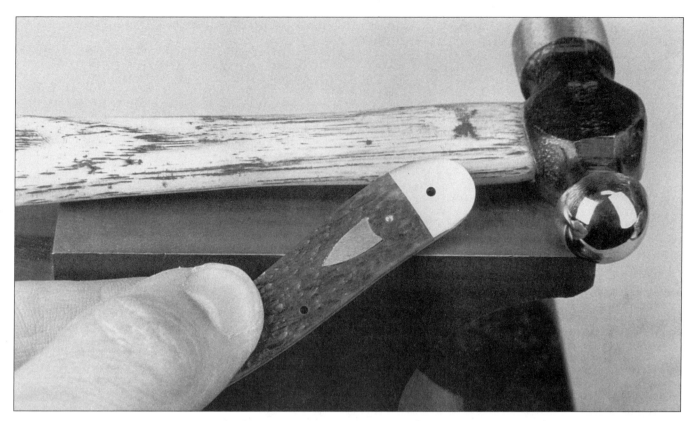

*Grind the shield smooth with 400 grit paper on the disc grinder.*

*Turn the ends of the pivot pin blanks to 1/8" to fit the lathe collet.*

*Lathe turn the pins first to 1/8" blanks.*

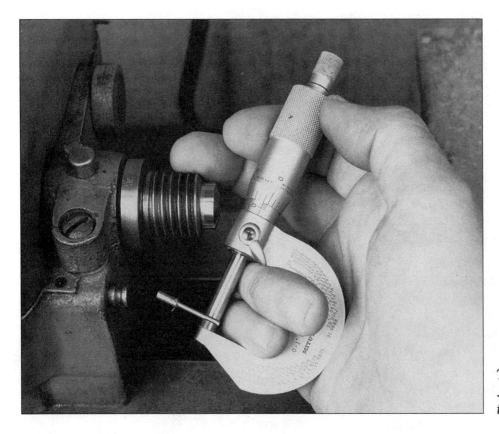

*The pin blanks are turned to .0760" to fit the pivot holes in the whittler's bolsters.*

*Relieve the inside of the handles .003" where the blade tangs would rub when closing the blades.*

less than 1/16 inch on each side of the liner and shield. Using 400 grit paper on the disc grinder, the cut ends of the pin are squared off. Using the steady and the flat face of the small hammer, give the pin a couple of light taps to swell it and lock the pin in the hole. The second pin is inserted, trimmed to length, and also locked in its hole. Now I use an air gun to carefully blow out the countersinks in the shield, and I give them a close inspection for any debris with the magnifier. If they are clean, use the peen end of the small hammer and carefully upset the pins. Watch the shield closely for any sign that the pins are distorting it and don't hit *anything* but the pins with the hammer!

The handle with shield in place now goes to the disc grinder. With 400 grit paper on the disc, grind the top of the shield smooth, rounding it from side to side. The shield should protrude about .010 inch above the surface of the jigged bone. It will be hand sanded and polished later.

The handle material pins are done next. Clean up the gold rod with 1500 grit paper and insert it in the hole. Cut the rod off so that there is 1/32 inch of the pin on each side of the handle. Square the ends of the pin on the disc grinder with 400 grit paper. Using the steady, lightly tap the pin with the flat face of the hammer to swell the pin in the hole. With the pin locked in the hole, turn the liner side of the handle up and *gently* peen the pin where it goes into the liner. The other three pins are done in the same way. I then put a tiny drop of Super Glue™ on each pin where it comes out of the jigged bone, and after it has had a second or so to wick in, wipe off the excess.

The handles now each have a piece of "handle" tape stuck on the bone side. Flatten the handles on the disc grinder with 120 grit paper. From there, they go to the drill press to have the handle material pins spun. With the 3/64-inch spinning tool in the drill chuck, apply a tiny bit of wax lubricant to the end of the tool. Set the drill press at a medium speed. Carefully spin the pins with light pressure until they are just below the bone surface. The other three pins are spun in the same manner, trying hard not to slip and mess up the handle material.

We'll turn the pivot pins on the lathe next. I started doing this, instead of using commercial pin stock, about five years ago. I always had problems with pivot pins showing in the polished bolster surface, either as a milky-looking spot, or as a halo where the pin didn't completely fill the countersink. Turning the pins from pieces of the same heat treated bar stock that the handles are made from solved the problem completely.

Begin by cutting 3/16 inch x 1 inch, 416 stainless bar into 3/16 inch long pieces. These are chucked in the lathe and a 1/8-inch section turned on the end. This area is gripped by a 1/8-inch lathe collet as the pieces are turned to make 1/8-inch pin blanks. I usually make up quite a few at a time. This is a good mindless job for heat-treating days. These blanks are turned to the pin size of the knife being assembled. In this case they are turned to .0760 inch for a snug fit in the whittler's pivot holes.

For the center pins on my bone and stag handled knives, I use heat-treated commercial pin stock. Because these pins don't have to "hide" as bolster pins do, regular drawn pin stock works fine.

Relieving the handle and spacer so the blade tangs don't rub when closed is an important operation on a custom folder. Brass and nickel silver liners are usually fairly forgiving when not relieved, but 416 stainless really requires it for the smooth operation of the blades.

I relieve my knives using a rotary jig in the milling machine. The jig was built of precision ground steel from an idea I got from "Lee" Thompson, who had built a similar jig. Basically, this jig holds the handle with the liner side up. It allows the milling cutter to remove about .003 inch from the handle in the area the blade tang would otherwise rub.

I first mark the handle where the relief needs to be cut. I also mark the location of the bottom of the spring. Then, the handle is clamped in the jig and the depth of cut is set to .003 inch. The relief is cut following the marks on the handle by moving the mill table and rotating the jig. The relieved areas on the spacer are cut the same as the handles, except that the spacer must be clamped more carefully because of its taper.

I used to cut these relieved areas with a sanding drum and my flex shaft tool. It did an acceptable job, but nowhere near as clean and neat as the mill can do.

After all the relieved areas are cut, remove the machining marks with a Cratex wheel in the flex shaft tool. Then hand sand the areas with a small rubber block and 1200 grit paper to level and smooth them.

The countersinks are now cut in the pivot holes. These help keep the pins from loosening when the knife is in service. I use a small rotary file with a "Christmas tree" shape and a 20 degree included angle, to cut these countersinks using the drill press. The depth of this cut should be about 1/32 inch. After all four pivot holes have been countersunk, I use a magnifier to check the cut. If the cut area is not perfectly smooth, I turn a cutter by hand, very lightly, to remove any roughness. I can't stress too much that if the countersunk area is not clean and smooth, the pivot pins will probably show after the bolster is polished.

Now put the two handles together, with pins in both pivot holes, and finish the bottom of the handle. First I check to see that the top of each blade is just "sunk"

*Countersink the pivot holes about 1/32". This will aid the pin's holding ability.*

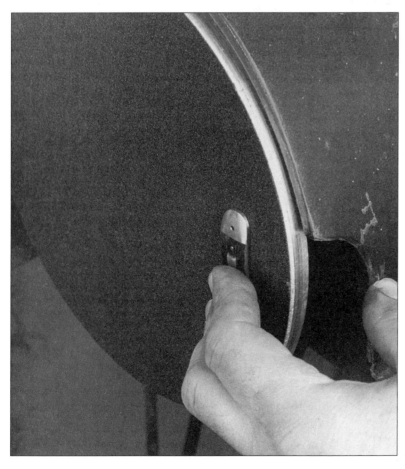

*Final finish the inside of the handles with 220 grit paper. Hand turn the disc for control.*

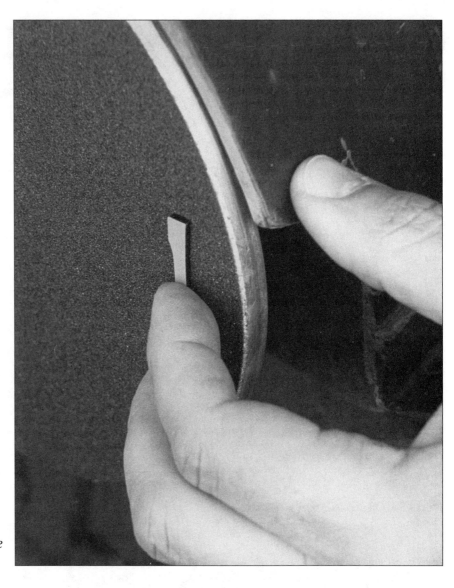

*Clean up both sides of the springs to remove any burrs.*

in the handle when in closed position. Then I use an 800 grit belt, on the 3-inch contact wheel of the small radius grinder, to smooth and perfect the profile of the bottom of the handle. The handles are next turned slightly sideways to round the outside corner of the bone and the bolster. While still pinned together, take the handles to the bench and sand the bottom of the handles with 1200 grit paper on a rubber block. Also sand the inside edge of the liners on the bottom of the handle to eliminate the sharp edge there.

Each handle is now given its final flattening on the disc grinder with 220 grit paper. Check carefully to see that all of the earlier coarse grit scratches are removed and the handles are perfectly flat.

The bevel is now cut that will hold the main blade straight at the front of the handle. Because the divider is .080 inch thick, each handle needs to be "out" .040 inch from the centerline of the knife. I take a piece of .040-inch-thick steel and fasten it to the rear of the handle with a 0-80 screw through the rear pivot hole. Then using the disc grinder with 220 grit paper, I gently press the handle to the rotating disc. Because the rear of the handle is held away from the disc, the disc will cut a slight bevel at the front bolster. Cut this bevel until it reaches the back of the front pivot hole and then stop. Both handles are

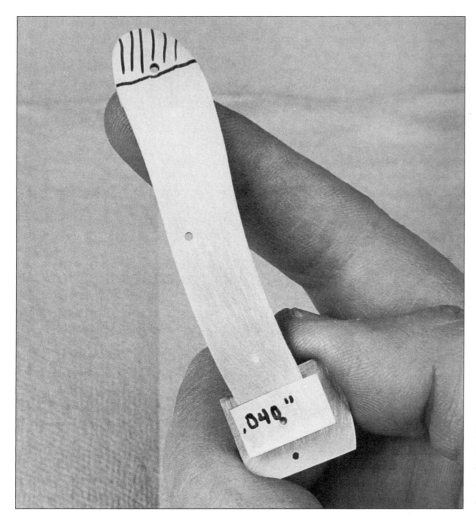

*Bevel the front (cross-hatched area) of each handle on the disc by fastening a .040" spacer to hold the rear of the handle away from the disc.*

beveled in the same fashion. We are *finally* ready to put the knife together!

## Assembly

Actually putting the knife together takes very little time once everything is prepared. Assembly is the step, however, where everything we've done can be ruined with a couple of hammer blows.

The way I set pivot pins is, to my knowledge, different from most other makers. Setting, or expanding the pins, as I do, is different from peening the pins. Peening, or cold heading, uses the *rounded* end of the hammer to flare only the end of the pin. To set pins I use the *flat* face of the hammer, to expand the entire pin. I think that expanding the pin has advantages, which I'll try to explain.

In order to open and close easily, the blade on a folder has to have a little clearance between it and the handles. If the handles were to be pressed together, the blade would bind. Most lockback knives have a bushing, a short cylindrical tube, that holds the handles apart and acts as a pivot bearing for the blade, however, most multi-blades don't use bushings, so the pin has to be tight enough in the handles to keep them apart. This is why the first few blows I use when setting pins are hard ones, meant to expand the pin through its entire length. I want the pin to swell, to tighten itself in the bolster. This, and the fact that the pin expands a little in the blade pivot hole, assures that the handles can not move. After the initial hard blows, I switch to lighter blows to fill each countersink and finish locking the pin in place.

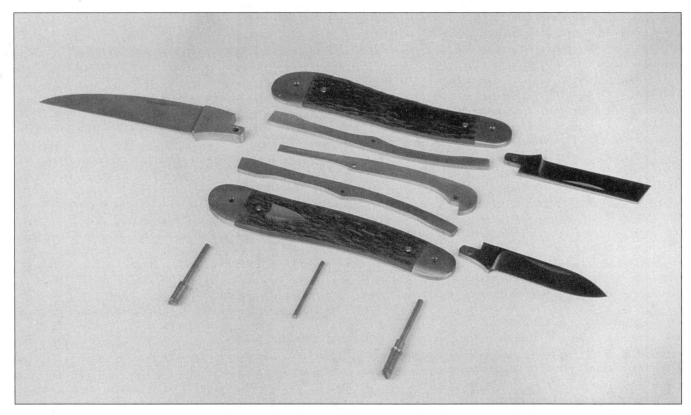

*Here are all the whittler parts laid out ready for final assembly.*

With the explanation of pin setting out of the way, we can concentrate on putting the knife together. Before I put anything together I first clean all the holes and pins using acetone. I think this helps the pins get a good grip inside the holes. Then I run my fingers over all the edges on the parts to check for any burrs that might hold anything apart. Any burrs are smoothed on the disc grinder with 1200 grit paper.

Assemble the handle parts stack on the center pin and hold it up to the light. No light should be seen between the handles, springs, and spacer. The small blades are put in place and the turned pin is carefully inserted. If it binds in the hole it may gall and stick. Now, take the knife to the spring press, compress the springs, and insert the turned pin after putting the main blade in place.

The pins are all cut off with the nippers, leaving about 3/32 inch protruding from the knife. The knife is next taken to the disc grinder to square the ends of the pins and bring them to the proper length.

Finding the proper length for the pins, before they are set, takes a little experience. If they are too long, they will tend to split as they are hammered and will leave a small crack when the pin is ground flush with the bolster. If the pin is too short, it will disappear into the countersink before filling it. Obviously a large countersink will use up a lot of material before filling.

Use 220 grit paper on the disc to grind the rear pin, leaving a little more than 1/16 inch sticking out of the bolster on each side. Grind the main blade pin to leave 1/16 inch protruding on each side. The pins must be ground perfectly square on their ends. Any other angle will cause them to cant to one side when struck by the hammer. I use 1200 grit paper to knock the sharp edge off the ends of the pins.

To assure that the blades will be free to move, I place .001 inch shim stock between the blade and the handle on one side of the knife. Two pieces of shim stock are used, one above the pin and one below. I hold these shims in place with a piece of masking tape. If one of these shims were to come out unnoticed while the pin was being hammered, it would lock the knife up. Use the air hose to blow out the countersinks and check with a magnifier to see that no little pieces of debris have found a home there.

On any multi-blade, I usually start with the rear pin. I place a spring clamp between the rear bolster and the center pin to make sure the handles and springs are held together. We are ready for the hammer.

Hammer in hand, hold the knife on the steady so that the pin is straight up and give the pin a good, solid, controlled blow. Make absolutely sure that the hammer face is square with the pin. I can't emphasize this too much. Now turn the knife over and do the same to the other end of the pin. Watch to see that the pin is still the same length on both ends. Check the knife to make sure it hasn't started to skew one way or the other. This will spoil the knife if left uncorrected. Assuming all is well, give each end of each pin two more slightly less vigorous whacks and check the alignment again. Now each end of the pin is hammered in turn, with fairly light blows, until the countersinks have been filled in. For insurance go back and give each end about ten more blows.

Pull the shims out and place a small drop of Break Free lubricant on the tang of each of the small blades. Check each of them for free operation. If a blade is tight, tapping on the pin on the other side of the knife will usually loosen the blade up.

The main blade pin is set next. Place the two shims as they were for the small blades and tape in place. The sequence of hammer blows is the same as it was for the other pin, except that these blows should not be quite as hard since this pin is not as long. After the hammer work is done, the shims are pulled out and the blade is lubricated. Check the blade for free movement and for any wiggle at the tip when the blade is closed. If the blade is a little loose it can be tightened up by placing a pin end on the steady and giving the other end one LIGHT tap with the

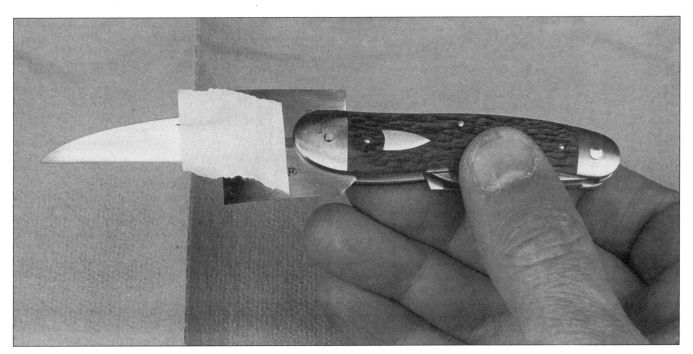

*The front pivot pin, with .001" shims in place, is ready to be set. The rear pin has already been set.*

Knifemaking with Terry Davis

*Tap the front pin very lightly to tighten the main blade.*

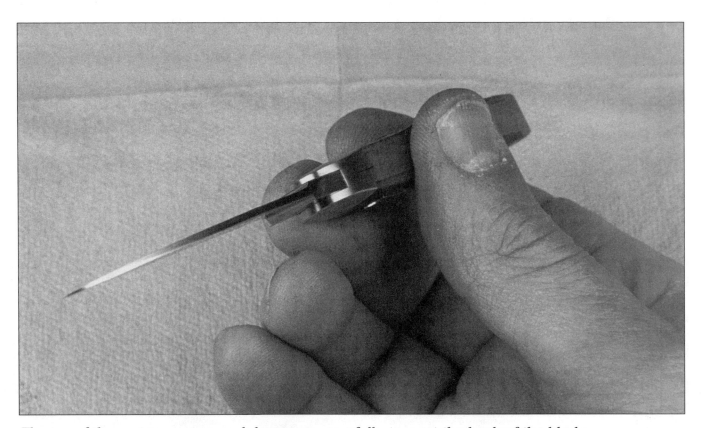

*The top of the springs are ground down very carefully to meet the back of the blade.*

hammer. Check the blade for movement and wiggle again. If it's still a little loose, go through the tap procedure again as many times as necessary. As long as this procedure is followed slowly, the blade won't tighten up too much. Once it gets too tight, about the only solution is to drill out the pin and start over again, which is not too much fun.

Next give the center pin a couple of good taps on each side. I put a spring clamp on each side of the center pin and after it has been set lightly, place a drop of Super Glue™ at each end of the pin.

Great! The knife is finally together. Take the knife to the belt grinder. Using a 400 grit belt, profile the top of the springs at the main and small blade ends to meet the top edge of the run up. Check the amount of the spring end that is above the back edge of the blade, close the blade, and grind the top profile down a little. This is repeated until the top of the spring meets the top edge of the run up on each blade.

Now, while the 400 grit belt is on the belt grinder, grind the pins flush to the bolsters. The short section of slack belt, below the platen, is used to grind a smooth contour on all the bolsters. Use a sharp 220 grit belt to grind each end of the center pin almost even with the bone. This must be done slowly so the pin doesn't overheat and burn the bone.

## *Polish and Sharpen*

This is really the home stretch. The knife will be done before we know it.

Spin the center pin first. I fold a paper towel to protect the knife from the drill press table and, holding the knife as immobile as possible, spin each end carefully. Any slips can scar the bone handle. Next, use 1200, then 1500, and finally 2000 grit paper to sand the shield with a hard rubber backing block. It will be polished later.

The bolsters are now sanded with 800 grit, then 15 micron, and finally 9 micron belts using the short slack belt section just below the platen on the belt grinder. All blades are closed to protect them. During this fine sanding, change the angle of the bolster to the belt periodically and check the bolsters to see that the previous grit scratches are removed. If the bolsters

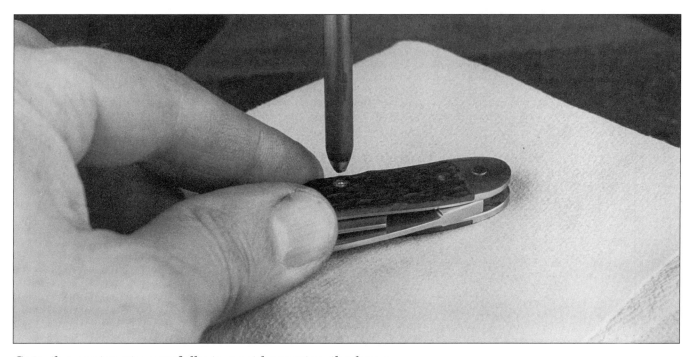

*Spin the center pin carefully to avoid scarring the bone.*

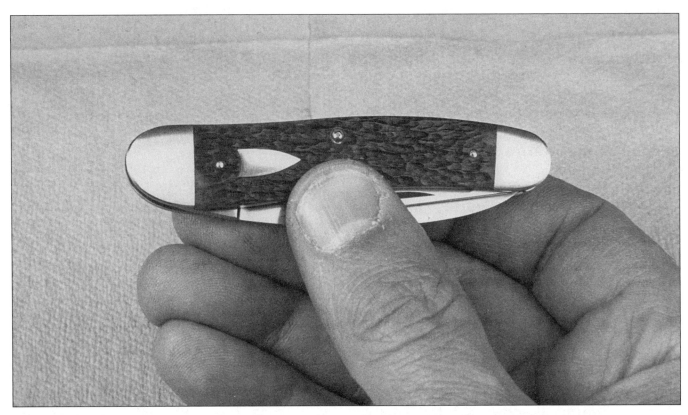
*Sand the bolsters with the belt grinder. The last belt used is a 9 micron belt.*

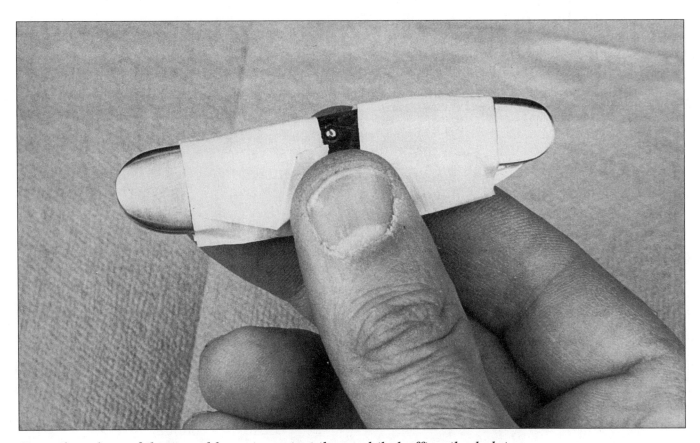
*Tape the edges of the jigged bone to protect them while buffing the bolsters.*

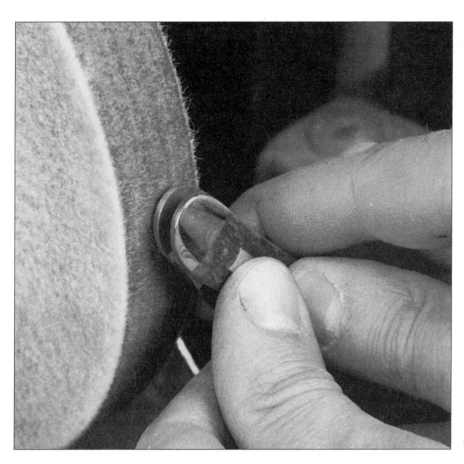

*Buff the edges of the bolsters by buffing from the bolster toward the jigged bone. Inattention will lead to the bone being buffed away.*

*Clean the buffing residue off with spray solvent.*

*Grind the blade swedges with a new 800 grit belt.*

are checked under a fluorescent light, the scratches can be seen easily. Every scratch looks like a ditch under fluorescent lighting.

Take the knife to the small radius grinder and with an 800 grit belt on the 3-inch wheel, round the edge between the top profile of the knife and the side very slightly. Then, using 1200 grit paper and a hard rubber backup block, hand sand the rounded edge of the handle all the way around both handles.

Use masking tape to protect the edge of the jigged bone handles on both ends of the knife. This will keep the bone from being touched while the bolsters are being buffed.

Buff the bolsters first with SS306 compound on a soft felt wheel. The bolsters should be buffed, changing the angle of buffing with each contact, until the sanding scratches are gone. Remove the tape and buff very carefully from the bolster toward the bone to remove any scratches at the edge of the bolster. If this is not done carefully the bone will be buffed away, leaving a raised bolster edge. Buff the handle *very* lightly with white Kocour compound on a 4-inch soft felt wheel. Heavy buffing will melt all the jigging cuts together and destroy the looks of the handle.

The shield is buffed lightly with red rouge on a 4-inch soft felt wheel. Excessive buffing will round the edges of the shield.

Finish the bolsters with green chrome rouge on a soft felt wheel. Doing the buffing under florescent lighting will help show up any scratches.

Before the knife is opened, I use Dayton 2X726B spray solvent to clean the buffing residue out of the knife innards and jigging, followed by a blowing-out with the air gun. This is a pretty nasty solvent, so do this in front of an exhaust fan or outdoors. When the solvent has dried, I put a small drop of Break Free on all the blade

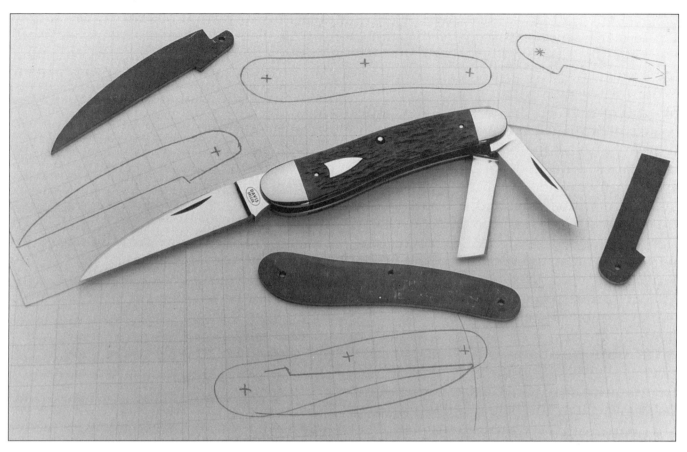

*From design on paper, to steel patterns, to a finished knife, you've seen it from start to finish.*

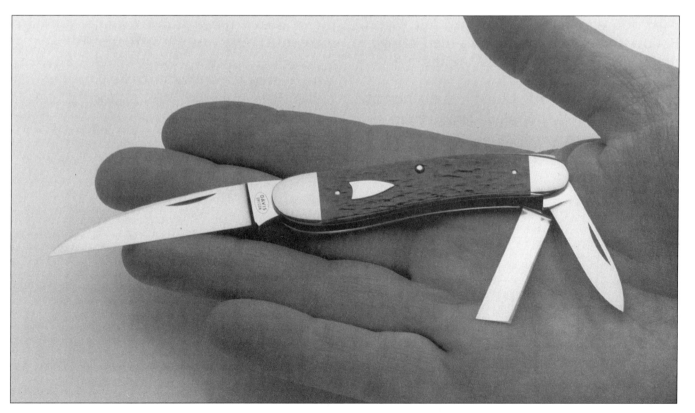

*A Wharncliffe Whittler and the hand it's meant to fit.*

tangs. I open all the blades and use a Q-Tip to clean any residue out of the knife. Wipe the blades off and we're ready to do the swedges.

Cut the swedges with a new 800 grit belt on the platen of the belt grinder. The Wharncliffe main blade has swedges running from above the nail nick to the tip of the blade. Take a deep breath, cut the swedge on one side, then try to match the starting point as the swedge is cut to the tip on the other side of the blade. (Getting the swedges to match is the interesting part.) Care must be taken, as the tip of the Wharncliffe blade is easily overheated. The swedges on the small blades are cut in the same fashion, except they are cut very lightly.

The knife is now sharpened. Use a worn 800 grit belt to grind the edge until it just forms a wire edge. Grind edge-down so the belt can't catch the blade. I take my time sharpening to get a nice, even edge. The extreme tip of straight-edge blades has to be watched very closely. It is easy to grind it away. Next, buff the edge until sharp with 555 compound on a hard felt wheel. Follow this by buffing with green chrome on a soft felt wheel. This makes a razor blade look dull!

Sharpen one blade at a time all the way through buffing the edge. Then, wrap the knife in padding and place it in the sanding vise. Using 1500 grit paper and a hard rubber block, sand each side of the blade from tang to tip in one stroke. This is a time for real concentration. The blade is very sharp. Do all the blades the same way—*carefully*.

The very last operation on the knife is to lightly grind the back of the knife, with the blades closed, with an 800 grit belt. The back of the knife should be a clean, ground surface. Then sand the back very lightly, using long lengthwise strokes, with 1500 grit paper on the hard rubber block.

The knife is done. It feels good in my hand and will be a good companion. Nothing equals the satisfaction from making a good pocketknife to carry and use, except maybe making another.

Gene and I both hope you've enjoyed this little knifemaking journey. It's been fun!

*The knife is finished and in my pocket and I want to say "thank you" to a couple of people before I forget.*

*The first thank you is to my wife, Cheyleen, who must have proof-read this a couple of dozen times. I don't know how she ever stood it, or me, for that matter.*

*A special thanks goes to "Gene" Shadley who offered me a chance to be a part of this book. He provided the spark to get it started and keep it going.*

# A Gallery of Multi-Blades

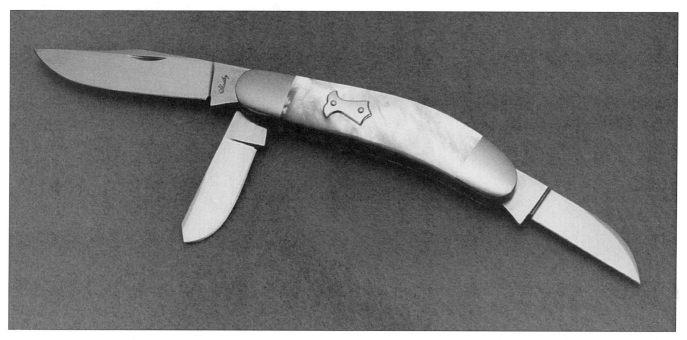

**Eugene Shadley:** *A three-blade sowbelly folder with a pearl handle. This is the knife that Gene used as an example in preparing this book's section on working with pearl handles.*

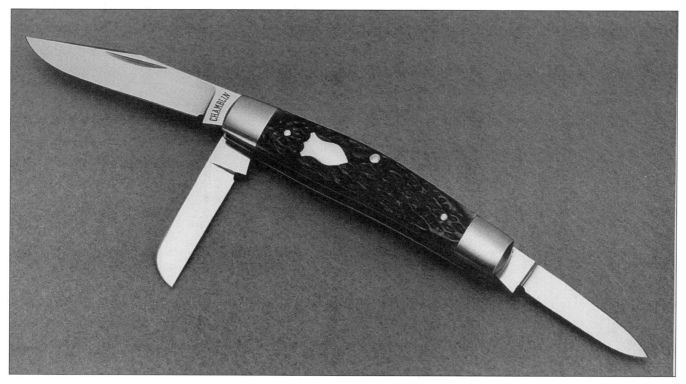

**Joel Chamblin:** *A three-blade stockman with a jigged bone handle.*

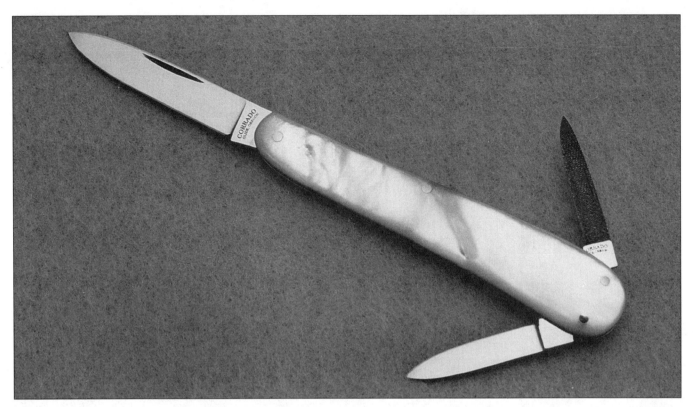

***Jim Corrado:*** *Two-blade lobster with file, 3-1/16" closed length. Mother-of-pearl handle, AEB-L stainless blades and titanium liners.*

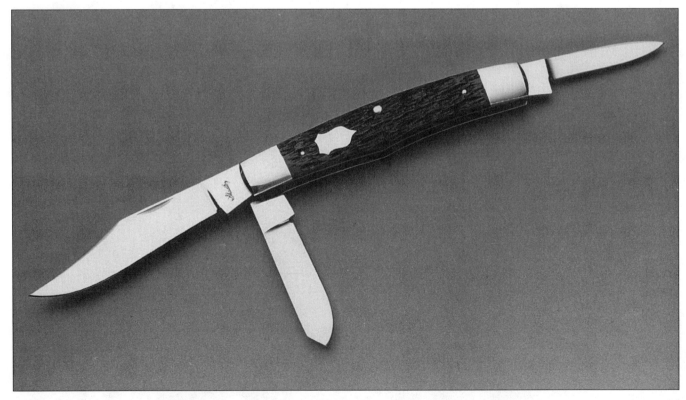

***Eugene Shadley:*** *Three-blade stockman with a jigged bone handle, 3-5/8" closed.*

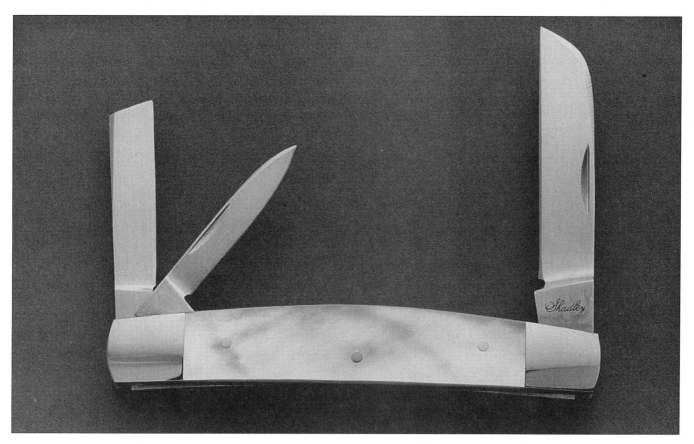

**Eugene Shadley:** *Three-blade Congress whittler with a pearl handle, 3-1/4" closed.*

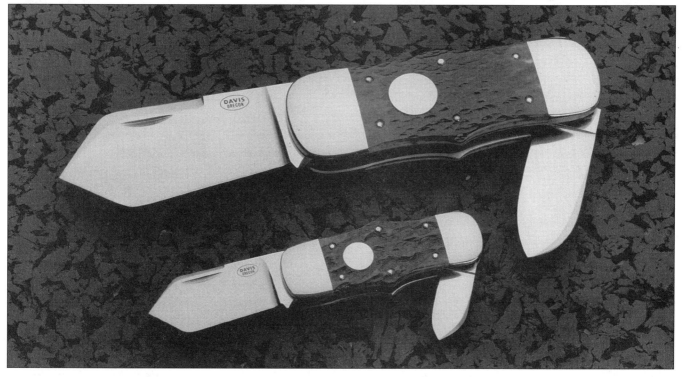

**Terry Davis:** *The regular Sunfish at the top is 4-1/4" closed; the Mini Sunfish is 2-1/2" closed. Both knives are in jigged bone with ATS-34 blades and 416 fittings.*

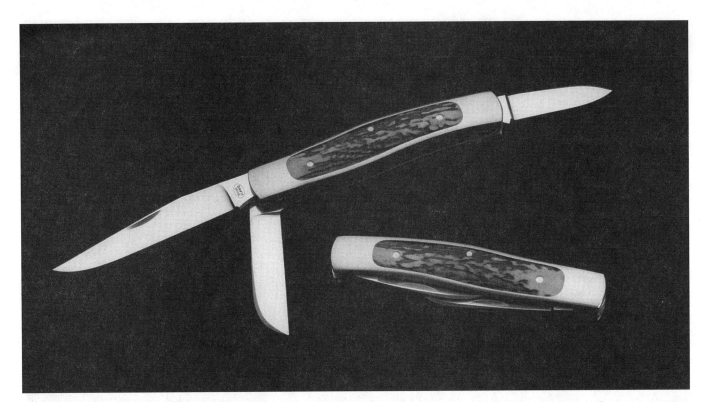

***Terry Davis:*** *These Interframe Stockman knives have stag inlays and are 3-5/8" closed. The blades are ATS-34 and all fittings are 416 stainless.*

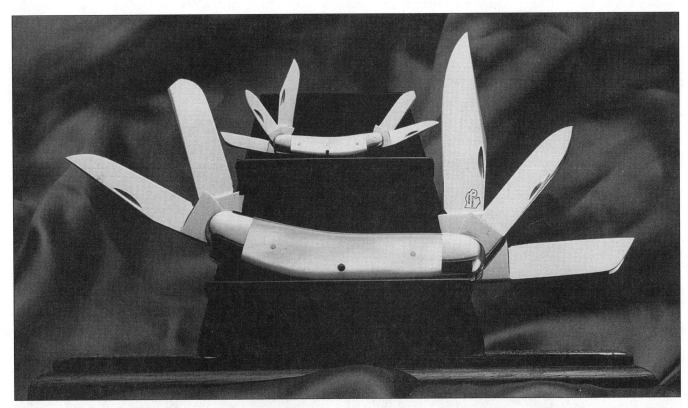

***Robert Enders:*** *Two five-blade sowbelly pocketknives, both with pearl handles, ATS-34 steel and nickel silver fittings. The bottom knife is standard size, 3-5/8" closed; the top knife is a 1-1/2" mini sowbelly.*

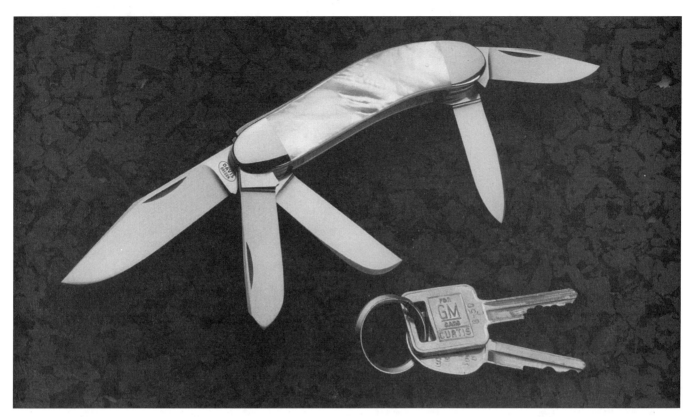

***Terry Davis:*** *This 3-1/4" closed length, 5-blade Sowbelly has white pearl handles with 416 fittings and ATS-34 blades.*

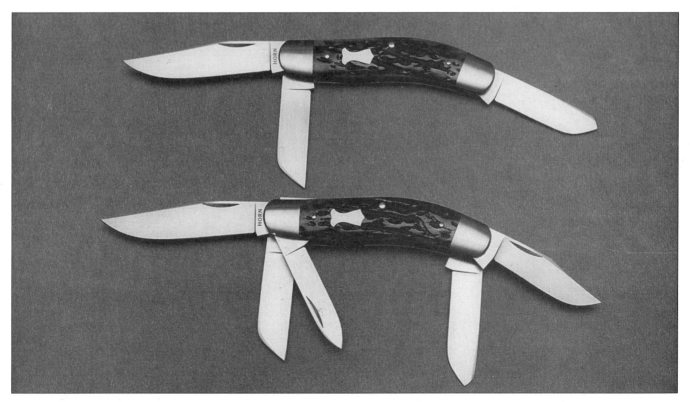

***Jess Horn:*** *Three- and five-blade sowbelly knives, both featuring ATS-34 blades and jigged bone handles. Closed length on five-blade 3-3/4".*

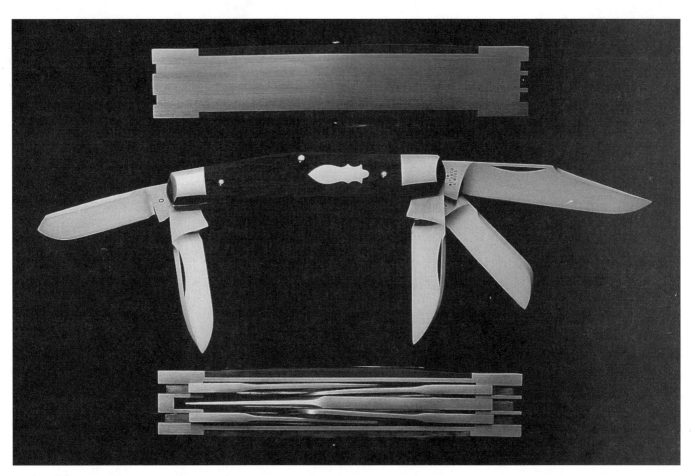

***Tony Bose:*** *Five-blade stockman with a jigged bone handle.*

***Mike Mercer:*** *A tiny six-blade Congress knife with a tortoise-shell handle.*

# Glossary

**Acetone:** A clear, extremely flammable solvent used to remove Super Glue™ and degrease knife parts before assembly.

**Anvil:** A block of steel used to hammer against.

**Assembly Block:** A piece of steel with holes drilled in it that holds pins while assembling the folder.

**Assembly Pin:** Temporary pin used to hold parts together while building the knife.

**Back:** The top of the blade, opposite the edge.

**Back Square:** The notched portion of the tang where the spring fits when the blade is in the open position.

**Belly:** The open side of the pocketknife handle.

**Bolster:** The heavier metal area at one or both ends of the frame or handle of a folder.

**Bolster Pin:** Pin used to hold bolsters together.

**Burr:** A ridge of metal left after drilling or machining.

**Center Divider:** A spacer that separates the springs and small blades on a whittler.

**Center Pin:** The pin in the center of the back of the frame or handle that allows the springs on a multi-blade knife to develop pressure to hold the blades open or closed.

**Choil:** Relief area between tang and cutting edge.

**Chuck:** A mechanical device that holds drills in a drill press or objects being turned in a lathe.

**Countersink:** The tool used to create a beveled opening at the end of a drilled hole that the peened head will fill so that it will not extend above the surrounding surface. Also, the opening itself. Also, to use the countersink tool.

**Cratex Wheel:** Impregnated rubber wheel used for polishing metal.

**Crinking:** Bending blades to allow room for the front ends of the blades to pass by each other without touching when the knife is closed.

**Crock Stick:** Round ceramic rod used for deburring or sharpening.

**Cutoff Pen Blade:** Also known as a coping blade. A blade design, usually small, that has the top and the edge of the blade parallel and the end sloping down to the tip at a sharp angle.

**Cutting Fluid:** A liquid, sometimes an oil, that helps a drill or other tool cut metal cleanly by lubricating and carrying off heat.

**Demagnetizer:** Device used to remove the magnetic field from steel.

**Dummy Pins:** Non-functional pins used for appearance only.

**Etch:** A mark created by acid or electrolyte and electric current.

**Flycutter:** A versatile tool holder used in a milling machine. It allows a simple toolbit to be used to level surfaces or cut special shapes, like nail nicks.

**Frame (or Handle):** Terms used by the authors to define the metal sides of the pocketknife. Also known as the "scale."

**Galling:** The tendency of two pieces of metal to stick to each other and, when moved, tear the surface of the softer piece.

**Grit:** A measure of the size of the abrasive particles on a grinding belt or sheet of sandpaper. As the numbers get larger, the "rocks" get smaller—400 grit is much finer than 60 grit.

**Heat Treat:** To change the crystalline structure of a metal by heating it to a certain temperature. This heating changes the properties of the metal, making it harder or softer, tougher or more brittle.

**Heat-Treating Foil:** A stainless steel foil used to enclose pieces of metal during heat treatment, protecting them from oxidation at high temperatures.

**Heat-Treating Furnace:** A firebrick-lined chamber capable of controlled heating to at least 2000°F.

**Jigged Bone:** Knife handle material of bone that has indentations cut into the entire outside surface. Usually dyed in shades of brown, red or green.

**Jig Plate:** Metal plate used to hold parts in position for machining.

**Kick:** Area at the bottom front of the blade tang that keeps the blade from hitting the spring when the blade is in closed position.

**Klingspor:** One of the manufacturers of grinding belts; Hermes is another.

**Layout Fluid:** A blue or red liquid that brushes onto a piece of metal and dries to a smooth surface that shows scribe lines very clearly.

**Liner:** The thin metal on the sides of a pocket-knife; when combined with the bolster it forms the frame or handle.

**Master Blade (also Main Blade):** The largest blade in a folder.

**Master End:** The main/master blade is located at this end of the folder.

**Micarta™:** A trademark of Westinghouse that covers phenolic resin reinforced with layers of paper or cloth. Available in many colors, in sheets or rounds.

**Milling Machine:** A machine tool that uses a movable table to hold the workpiece and move it past a turning cutter. Very precise work is possible, but unless the machine is computer controlled, precision is dependent on the operator.

**Moto-Tool:** A small, hand-held grinding/polishing tool.

**Nail Nick:** A crescent-shaped cut in a blade allowing use of a fingernail to open the blade of a folder.

**Norton Hogger Belt:** A cloth-backed ceramic belt; long-lasting.

**Pakkawood:** A manufactured wood product made of layers of veneer glued under pressure, much like plywood. Very dense, hard and flat. Comes in many colors.

**Pattern:** Term used to describe the metal shapes used to scribe parts and guide drill bits allowing very precise duplicate parts to be made.

**Peen:** Using the "ball peen" end of a hammer, to swell the end of a pin in assembling a knife. Also known as "cold heading."

**Pen Blade:** A small spear blade in a multi-blade knife.

**Pin Spinning:** The process used to dome the head on a pin that creates a burr on the end of the pin, holding the handle material in place.

**Pivot Pin:** The pin in the end of a pocketknife that the blades pivot on when they open and close.

**Profiling:** Grinding parts to the scribed shape.

**Quench:** Cooling a piece of hardenable steel from above its critical temperature at a fast enough rate to cause it to remain in a hardened state. Depending on the steel type, this may require the use of moving air, oil or water.

**Sanding Plate:** A piece of steel, ground flat, to which sandpaper can glued. Used to deburr metal parts.

**Scotch Brite™:** Abrasive pads that come in several grit sizes.

**Scribe:** Using a sharp, pointed instrument to scratch a mark on steel.

**Setup Pin:** A pin to hold knife parts for fitting and to assemble the knife temporarily to check function.

**Sheepsfoot Blade:** A blade design featuring a straight cutting edge, with the top of the blade curving sharply down to meet the edge at the point.

**Slack Belt Finishing:** Using the portion of the belt that has no support behind it, allowing more flex in the belt and producing more flowing contours on the finished metal piece.

**Spacer:** The strip of metal between each layer of springs and blades in a multi-blade knife.

**Spey Blade:** Thin, sharp blade with a rounded point.

**Spring:** The piece of metal that bears on the tang of the blade in a folding knife. This part must be tempered and shaped correctly to allow it to flex and still keep pressure on the tang.

**Spring Press:** The tool that pushes springs up, allowing the knife to be assembled.

**Stag:** Antler from European or Asian elk or deer.

**Steady:** A small anvil used in cutlery work.

**Steady Rest:** An attachment used when grinding parts that holds them at a specific angle; also called a work rest or tool rest.

**Steel Types**

**A-2:** An air-hardened steel used in this book to make working patterns. Although an excellent blade steel, it has only 5 percent chromium and so is not completely stainless. It makes good patterns because it shows very little distortion when heat-treated.

**O-1:** An oil-hardened carbon steel available in precision-ground bars. Easy to harden and temper without specialized equipment.

**L-6:** Carbon steel available mostly as salvage from sawmill blades. Very easy to harden and temper and very tough and flexible because of its high nickel content. Oil hardened.

**ATS-34/154CM:** High-quality stainless blade steels that are almost identical in composition. A high chromium content makes these steels very difficult to heat treat because they must remain at high temperatures for long periods. Both require a subzero quench to completely harden.

**416:** Stainless steel used for knife parts other than blades and springs. Will harden slightly, which improves stain resistance. The main advantage is its close color match with ATS-34.

**410:** Very close in composition to 416. Mostly used for spacers and shields in multi-blades because it comes in sheet form.

**Mild Steel:** Low carbon non-hardenable steel, this is the common steel used for everyday products like nails and tin cans.

**Stress Relieve:** Using a controlled heating sequence to remove internal stresses caused by machining and grinding of blades and springs. This helps eliminate warpage when hardening.

**Sub-zero Freezer:** Freezer capable of reaching at least -100°F.

**Swedge:** Relief ground on the top of a blade for clearance.

**Tang:** The rear end of a folding knife blade.

**Temper:** To soften a hardened piece of steel to make it tougher and more suitable for its desired use.

**Test Coupon:** Term used in this book to describe a small strip of blade steel used for quality control during heat treating.

**Thin Film Resist:** A liquid that dries to form a film and is used to create a stencil to acid etch markings on blades.

**3M™ Apex Belt:** Cloth-backed belts with pyramid-shaped abrasives.

**3M™ Micro Film Belt:** Plastic film belts good for slack belt sanding.

**Walk Area:** The area of the end of the spring that the back square of the blade rubs on.

**Wharncliffe:** A blade with a straight cutting edge and a gradual taper from the tang area to the tip.

**Whittler:** A pocketknife design in which both of the knife's springs bear on the main blade tang and each spring bears on a small blade tang at the other end of the knife.

**Woodruff Key Cutter:** A wheel-shaped milling cutter that is used in this book to cut the back square area of the blades.

# Addresses of Multi-Blade Makers Featured in this Book

Eugene Shadley
645 Norway Drive
Bovey, MN 55709-9508

Terry Davis
Dredge Lane Road, Box 111
Sumpter, OR 97877

Tony Bose
7252 North County Road, 300E
Shelburn, IN 47879

Jim Corrado
2915 Cavitt Creek Road
Glide, OR 97443

Robert Enders
3028 White Road
Cement City, MI 49233

Jess Horn
87481 Rhodowood Drive
Florence, OR 97439

Mike Mercer
149 North Waynesville Road
Lebanon, OH 45036

Joel Chamblin
Route 1, Box 98
Concord, GA 30206

# List of Suppliers

**Admiral Steel Co.**
4152 West 123rd Street
Chicago (Alsip), IL 60658-1869
800-232-7055
(ATS-34 steel)

**Kodak Marketing Center**
1133 Ave. of the Americas
New York, NY 10036
212-930-8000
(Kodak Thin Film Resist marking supplies)

**Chris Peterson**
Box 143, 2175 West Rockyford
Salina, UT 84654
801-529-7194
(damascus steel billets)

**Barry & Sewall**
2001 NE Broadway
Minneapolis, MN 55413
800-328-5486
(Loctite Depend®)

**Koval Knives**
5819 Zarley Street
New Albany, OH 43054
800-556-4837 (out of state);
614-855-0777 (in Ohio)
(Micro-lathe)

**POP Knives**
103 Oak Street
Washington, GA 30673
706-678-2729
(2x72 belts, Klingspor sandpaper)

**Paul Bos Heat Treating**
1900 Weld Boulevard
El Cajon, CA 92020
619-562-2370
(Heat treating)

**MSC Industrial Supply Co.**
151 Sunnyside Boulevard
Plainview, NY 11803-9915
800-645-7270
(bits, end mills, surface-grinding wheels)

**R. E. Roberts**
PO Box 3507
Gastonia, NC 28052
(handmade steadies)

**Brownells, Inc**
200 South Front Street
Montezuma, IA 50171
515-623-5401
(555 Polishing compound, Pana Vises)

**Masecraft Supply Co.**
170 Research Parkway #3, PO Box 423
Meriden CT 06450
203-238-3049
(pearl)

**Cindy Sage**
1106 Northwest 3rd Avenue
Grand Rapids, MN 55744
218-326-6339
(cloth slip covers)

**Fry Steel Company**
PO Box 3585
Santa Fe Springs, CA 90670-1585
310-802-2721
(416 stainless bar and pin stock)

**Micro-Mark**
340 Snyder Ave.
Berkely Heights, NJ 07922
800-225-1066
(small tools of all kinds)

**Sheffield Knifemakers Supply**
PO Box 141
DeLand, FL 32721
904-775-6453
(ATS-34 steel, 416 pin stock)

**Harper Manufacturing Co.**
3050 Westwood Drive, #B-5
Las Vegas, NV 89109
702-738-8467
(steel marking stamps)

**Mother of Pearl Co.**
293 Belden Circle
Franklin, NC 28734-0445
704-524-6842
(pearl and shell)

**Texas Knifemaker's Supply**
10649 Haddington #188
Houston, TX 77043
713-461-8632
(304 stainless sheet stock, heat treating)

**Hoover & Strong**
10700 Trade Road
Richmond VA 23236-3000
800-794-3700
(gold and silver)

**Natural Products Co.**
266 West 37th St.
New York, NY 10018
800-789-STAG
(white bone, handle materials)

**Tru Grit**
760 E. Francis Street, Suite N,
Ontario, CA 91761
909-923-41163
(2x72 belts)

**K & G Finishing Supply**
PO Box 458
Lakeside, AZ 85929-0485
520-537-8877
(410 stainless sheet stock)

**Northwest Knife Supply**
621 Fawn Ridge Drive
Oakland, OR 97462
541-459-2216
(etching machine, Klingspor sandpaper)

**Weaver Leather**
PO Box 68, 7540 C.R. 201
Mt. Hope, OH 44660
800-WEAVER-1
(leather dye)